D1602745

How to Become a Professional Engineer

The Road to Registration

John D. Constance
Engineering Registration Consultant
Registered Professional Engineer, New York and New Jersey
Certificate of Verification (Records Verification Program)

Fourth Edition

McGraw-Hill Book Company
New York St. Louis San Francisco Auckland
Bogotá Düsseldorf London Madrid Mexico
Milan Montreal New Delhi Panama
Paris São Paulo Singapore
Sydney Tokyo Toronto

Library of Congress Cataloging-in-Publication Data

Constance, John Dennis, date.
How to become a professional engineer: the road to registration
John D. Constance.—4th ed.
p. cm.
Bibliography: p.
Includes index.
ISBN 0-07-012468-X
1. Engineering—Vocational guidance. I. Title.
TA157.C6 1988
620′.0023—dc19 87-16837
CIP

1234567890 BKP/BKP 893210987

ISBN 0-07-012468-X

The editors for this book were Betty Sun and Nancy Young, the designer was Naomi Auerbach, and the production supervisor was Dianne L. Walber. It was set in Century Schoolbook by Byrd Press.

Printed and bound by The Book Press.

Subtitle added for this edition.

To my wife Cecilia
whose cooperation, encouragement,
and sacrifices, since this book's inception,
have made its fruition possible.

Contents

Prologue

Steps on the Road to Registration

1. ABET Engineering Education or Board Approved Equivalent
2. Examination in Fundamentals of Engineering (FE)
3. EIT Certification
4. Approved Engineering Experience
5. Examination in Principles and Practice (PE)
6. Licensure/Registration

Preface

The increase in engineering activity throughout the country in the years following World War II has stimulated interest in the state registration laws that regulate the practice of professional engineering in all of its manifestations and in the administration and enforcement of these laws. Existing laws are continually being amended and new statutes are being enacted to replace inadequate laws to keep up with the times.

There is a trend that has manifested itself by placing "public" members on boards of registration. These individuals are not engineers nor are they technically inclined in engineering matters. There was some resistance initially to having public members on the boards, particularly with regard to giving each public member a vote on the technical qualifications of applicants; however, opposition has softened considerably as experience has indicated that these members will not be authorized to invade technical areas that are beyond their competence. Time has borne out the fact that public members on boards are good for the engineering profession as they have shown that there is nothing secret or sinister in the way boards handle their business. Public board members have proven to be good allies in helping the public understand the true purpose and public benefit of the registration laws. Public members primarily serve as a conduit for the public to the boards. This has helped to provide the public with knowledge of the boards' duties and has also assisted in providing the public with a greater degree of confidence in the registration process.

State boards have worked toward developing their own rules of professional conduct and have taken on the added burden of policing the profession for both legal and ethical transgressions in order to protect the public health, safety, and welfare. State registration laws have been amended to clearly give boards the authority to adopt binding rules of professional conduct and to demonstrate that the violation of such rules may be grounds for suspension or revocation of the license of the individual involved; therefore, it has become highly desirable for the state registration laws to give undoubted authority to the state boards to adopt rules of professional conduct, and for those board rules to clearly spell out the activities that are forbidden to licensees in order to protect the public interest.

Concurrently, the registration laws have been strengthened by increasing the requirements for registration. There is a determination through state legislation to provide higher quality of service to the public by this increase and by more aggressive enforcement of the registration laws.

Greater uniformity in the application of the registration laws has been another trend. As more and more professional engineers move across state lines to new employment and consulting situations, they will need to obtain registration in their new states with a minimum of effort. The National Council of Engineering Examiners (NCEE) has a ready-made Records Verifications Program that minimizes the effort and facilitates interstate registration.

Between now and the end of this century, society's demands on engineering will range from providing for interplanetary exploration and travel to solving many earthly problems: energy conservation, pollution control, transportation, housing, preservation of the environment, and others arising from the population explosion. These demands are limited only by our imagination, hopes, and technical and financial resources. They can be met only by the engineering team of engineer-technologist-technician becoming a broad continuum of mathematician-scientist-engineer-technologist-technician. They will best be met by men and women committed to a better world.

Engineers' responses to the previous editions of this book have been most gratifying and encouraging. The book continues to provide guidance and direction to the engineering student, as well as to the graduate who is already in practice, in setting a course that will lead to a successful professional career. Its purpose is to alert them to the importance and necessity as well as the long-range benefits of Engineers' Registration.

For the person who wants to know more about the reasons why Engineers' Registration is here to stay and grow, for the woman or man who wants to know more about requirements and procedures for legal registration, this book provides the necessary information and guidance. The chapter on qualifying experience is most helpful for orientation. Engineers who want to know how to document their experience records for the greatest effect, value, and credit will find the answers herein.

Most engineers are too modest in documenting their experience in the application, and they thereby fail to realize the full effective value of their hard-earned experience. Why hide your light under a bushel? In evaluating your experience record, your board of registration must reach its decision by what you have written into the record, which must give the complete story, leaving no doubt in the minds of the board members about the exact nature, quality, quantity, and character of your experience.

I believe this revised edition is a definite improvement over previous editions. It has been made more useful as a ready instrument to attain licensure. The glossary of terms used in registration procedures has been updated and improved, as have the suggestions for an engineering experience journal, which will be found useful for writing acceptable job experience descriptions when the time comes. An up-to-date definition of "responsible charge" clears up this often misunderstood expression. To help the uninitiated get started, a useful first-step self-evaluation checklist is included as a time saver as well as a money saver when determining one's qualifications.

The chapter on the written examination has been completely rewritten to reflect the NCEE national format. In addition, a discussion on the use of objective-type examination versus subjective-type examination has been included. The propensity toward the increased use of the objective examination format is highlighted.

Interest in the various modes of preparation for the written examination has grown considerably; therefore, the pros and cons of each mode—college texts, study guides, tutoring, refresher and correspondence courses, and videotaped courses—are discussed.

In addition to the above, improvements to the discussion of the following subjects have been included in this edition:

Experience Checklist
Application Checklist
Recommended Reference Texts and Study Guides
Qualifications for Registration
Classification of Applicants
Applicants with Foreign Degrees
Registration by Endorsement
Use of the PE Engineer's Seal
What Makes a Professional Climate?
The Licensing Procedure
Bibliography

The following material has been deleted from the previous edition:

How State Examination Procedures Compare

Licensure Through Eminence

Oral Examinations

The interested engineer seeking registration may have questions on these subjects. Local boards can supply information about the state registration laws of interest. To help in this area, addresses of state registration boards have been included in the Appendix.

Since all state boards now administer the NCEE examinations in the Fundamentals of Engineering (FE) and the Principles and Practice of Engineering (PE), any question about their administration should be directed to the board of interest. Questions about examination content should be directed to NCEE, since NCEE now develops the examination content for the FE and PE examinations.

The experience and application checklists, in their improved format, will continue to provide a strong "assist" to the applicant. The engineer of foreign birth and foreign experience and the engineering graduate will find the credentials list used by the New York State applicants of inestimable help.

The sample experience list has been retained and expanded to cover additional disciplines, and an in-depth experience record sample has been added, since the list has been an invaluable tool in the important matter of documenting an applicant's experience. For the applicant who may have a part of the written examination waived if a well-documented and detailed record is submitted, these tested samples are especially offered. Herein, task identification is used as the basis for describing qualifying experience which meets the standards for experience.

The section on special study aids has been reworked. Also, a discussion on multiple registration through NCEE's Records Verification Program has been added.

John D. Constance, PE

How To Use This Book

This book is intended to help both the young engineering student and the experienced graduate engineer to set a course that will ultimately lead to a successful professional career. Together with the older engineer who has spent a number of years in engineering work, these younger women and men, who look to Engineers' Registration as the legal manifestation of their professional careers, will receive help and direction.

At the very outset, the book outlines the various steps that license-minded engineers should follow in developing professional careers. The first chapter reflects much of the thinking of educators and industry today, which stresses the need for teamwork to answer our country's call for high-caliber engineering talent. The first few years after graduation are the most important to the young engineer. This is the critical period which spells either future success or failure in professional life. After "commencement," ambitions are no longer just dreams; they may become reality if the individual can see them through.

What is professional engineering and what is the purpose of Engineers' Registration? Chapters 2 and 3 attempt to answer these questions. Often considered a luxury by graduating engineers, the "PE" is integral to career development, to the integrity of our profession, and to areas of public safety affected by the practice of engineering. In a profession that essentially regulates itself through state boards, professional standards of competence and ethics are necessary to maintain public confidence and trust. Public trust is essential to the future of our profession.

Registration is a profession's tool to protect the public interest in matters relating to the engineering field. It is for this reason that state legislatures across the country originally passed registration legislation and why these laws continue to be in effect today. Those who are not registered but publicly call themselves "engineers," or use the term "engineering" to describe their work, are in violation of the state statutes. Chapter 4 presents a summary of state registration laws.

The presence of legislation doesn't reflect a lack of confidence among those currently working in industry without PE licenses. Many depend on the "Industrial Exemption" provisions to bypass licensure, but some states have removed such an exemption and other states are moving in that direction; thus, it behooves the engineer in industry who is practicing without a license to remain alert to the changing conditions. Many are highly competent individuals with outstanding credentials. But their continuing success in the engineering field without licenses should not imply that registration is not needed or desired by U.S. companies.

Many students plan to work for companies in industry that do not require them to be licensed at present. Why should they prepare and sit for the FE, work for years to gain qualifying experience, and then sit for the PE part of the exam? To them it is an inconvenience and not worth the effort. But this kind of thinking is short-sighted.

An engineering student should be encouraged to ask these questions: "Am I planning to embark on an engineering career that I expect to last 40 to 50 years? Is it possible that the PE license will become a requirement in the area of practice in which I am interested? Might my career goals as an engineer change? How about the strong possibility of the elimination of the industrial exemption within the near future in every state?" Most who do pursue eventual licensure successfully do so to provide career insurance and fulfillment of personal satisfaction. Chapter 5 describes the Engineer-in-Training (EIT) program, which is one route an engineering student can follow toward eventual licensure.

Requirements for registration are well documented and explained in Chapter 6. Classification of applicants and a self-evaluation checklist are offered in the Appendix to help the interested engineer to get started on the right path to licensure, providing answers to questions about qualifications and requirements.

What is qualifying experience? Chapter 7 discusses the various classifications in which an engineer's experience places him or her. This is very helpful information, and the engineer who wants to know more about how to write an experience record for greatest weight and value will find many of the answers here. The relation of engineering experience to licensure is given in-depth coverage. Experience and application checklists are given in the Appendix.

Chapter 8 provides in-depth coverage of the criteria used by boards of registration when evaluating professional and subprofessional experience for a number of engineering disciplines, including chemical, civil/sanitary/structural, electrical, and mechanical engineering. When documenting experience, the candidate can put these criteria to good use. Merely because one is a chemical engineer does not mean that the data in the other discipline listings can't be applied to the experience list. Engineers are encouraged to do just that.

Problems encountered by boards of registration in the evaluation of experience records are given in Chapter 9. Candidates should take heed of this discussion. Boards must reach their decision on the basis of what is written in the application form and especially by how well the evidence has been documented. They should not be required to read between the lines or go by mere hearsay. Checklists of personal and professional questions, records of active practice, and the details of preparation material to include when filing your application are highlighted. Also included are application materials required, together with reference forms. Of special interest are samples of tested experience records found in the Appendix.

In Chapter 10, the philosophy and purpose of the written examination are discussed, along with its requirements, how you can prepare for it, and how you can pass it the first time.

Both parts of the NCEE examination, Part A, Fundamentals of Engineering (FE) and Part B, Principles and Practice of Engineering (PE) are discussed in detail: the method of grading used, hints on performance, degree of preparation needed, time allotment for the exam, and general treatment overall. A detailed discussion is included about the method of grading by NCEE. Suggestions are offered for guarding against emotional block during the exam. What is permitted in the exam room is listed, although you will be given specific instructions by your board before you sit for the exam. The use of objective examinations and questions are discussed to enlighten the examinee about possible future examination format changes.

In Chapter 11 you are told what information to gather and how to get organized to prepare for the written examination. Strategies to use for the FE and PE parts of the exam are spelled out for you. An examination checklist is included, together with a list of organizations to write to for information on study aids, refresher courses, and videotaped courses.

Hints on how to select refresher courses are included in Chapter 12. Prerequisites, course organization, and individual course instructors are watchwords to remember. Before you put down your course fee, be sure you qualify to take the examination, or you may waste time and money.

How you obtain multiple registration and temporary practice permission in another state and how you obtain a Certificate of Verification are detailed in Chapter 13. Problems and pitfalls to avoid when seeking licensure in another state, the affixing of your seal to engineering documents, government-as-client limitations, and corporate practice limitations are additional subjects included.

Chapter 14 explains the reasons for and use of the engineer's seal. Certification of documents versus plan stamping is discussed as well, in an attempt to clear the air on terminology and application.

Definitions of various engineering disciplines are given in the Appendix to help the applicant write up an experience record. All the definitions should be studied for cross-referencing of experience.

Answers to many questions asked by applicants are included in the glossary of terms, found in the Appendix. This compendium offers quick and simplified answers, free of the legal jargon often found in the wording of state registration laws.

Oregon laws affecting the use of engineering titles and the overseas program for testing offered by the Minnesota Board are also included in the Appendix.

The Bibliography should be referred to for a study of Engineers' Registration (licensure) and other areas of general interest and importance.

Acknowledgments

The author gratefully acknowledges the help and cooperation he received from the publishers and editors of the published material referred to or presented below:

"What Makes a "Professional" Climate?" reprinted with permission from *Chemical Engineering* (December 5, 1977) McGraw-Hill, Inc., New York.

"Glossary for Professional Engineer Candidates" reprinted with permission from *Chemical Engineering* (July 1, 1968), McGraw-Hill, Inc., New York.

"Setting Up the Engineering Team" reprinted by permission from *Machine Design*, (September 11, 1980), Penton Publishing Company, Cleveland, Ohio.

The author also acknowledges with thanks the permission granted him by the National Council of Engineering Examiners (NCEE) to include the latest examination specifications for the Fundamentals of Engineering (FE) and Principles and Practice of Engineering (PE) exams included in Chapter 10.

The author is indebted to the publishers and editors of the following technical publications for their farsightedness and encouragement in providing space to help tell the story of professional registration: *Power Engineering*, *Chemical Engineering*, *Machine Design*, *Mechanical Engineering*, *Tau Beta Pi Bent*, *New Engineer*, *Heating Piping and Air Conditioning*, *Design News*, *ISA Journal*, *Actual Specifying Engineer*, *Building Systems Design*, *Heating and Ventilating*, *Consulting Engineer*, *Power*, *New York Professional Engineer*, *Petroleum Refiner*, and *American Engineer*. Lastly, the author gives heartfelt thanks to his many readers who have motivated him to provide them with a ready means for satisfying an existing need for licensure and who have helped formulate new ideas for the continued improvement of this work, truly a labor of love.

John D. Constance, PE

ABOUT THE AUTHOR

For more than 40 years John Constance has been helping engineers to prepare for professional registration, having given numerous P.E. refresher courses for engineering societies (ASME, ASCE, and IEEE) as well as such firms as M. W. Kellogg, Parsons-Jurden Corporation, and Stone and Webster. The author of more than 100 professional articles and two other books on engineering registration (all McGraw-Hill, *Mechanical Engineering for Professional Engineers' Examinations*, 4/e, and *Electrical Engineering for Professional Engineers' Examinations*, 4/e), Constance is a contributing editor to *The Military Engineer* and a registration consultant to the U.S. Army Corps of Engineers.

1

DEVELOPING A PROFESSIONAL CAREER

Before entering into the subject of developing a professional career, let us define "profession" itself. Professor W. C. Wickenden, in *The Second Mile*[16]* wrote:

> If one searches the authorities for definitions of a profession, he will probably find four kinds. One is likely to hold that the determining quality is *attitude of mind*, that an altruistic motive can lift any honorable calling to the professional level. A second may say that it is a certain *kind of work*, one requiring special skill on a high intellectual plane. A third may state that it is a special *order in society*, as the bar, the bench or the clergy. Still others insist that no work can be professional without a *confidential relationship* between a client and his agent, as that of a patient to a physician, litigant to lawyer, and so on. None of these definitions is self-sufficient. Taken together, like the legs of a table, they give a profession a stable base of support.

What constitutes a professional? As we have seen, different people have defined the professional in different ways. Professor Wickenden says professionalism is a matter of spirit and scope. Vannevar Bush states that the professional should minister to the people with dignity and authority. Our Accreditation Board for Engineering and Technology (ABET) says it in another way indeed:

> It is impossible to over-emphasize the fact that technical competence is the necessary but not the *sufficient* condition for becoming a professional

* Superscript numbers refer to the bibliography at the end of the book.

engineer. There must be developed, in addition, certain personal characteristics, qualities of personality and character, which will establish unmistakably a feeling of trust, confidence, and loyalty in man-to-man relationships.

Some years ago, an editorial in *Civil Engineering* stated:

> The engineer who considers his calling to be a profession will be interested to learn the obligation he has assumed to justify so high a status. A professional man must, of course, possess special knowledge and skill and the ability to exercise them but, of equal or greater importance, he must possess altruistic characteristics and he must conform rigidly to a code of ethics of a high order.

The true professional should do a bit of soul-searching and ask, "Am I really professional in my attitude, in my relations with my employer, my employees, my clients, and with the public in general?"

How many present-day practicing engineers thought of engineering as a profession when they planned their preparation? Probably very few looked into the matter carefully and completely. Yet many of them have been active in one way or another—in society work and on committees, in considering ways and means of enhancing their professional status, and in helping the younger engineering graduates to develop professionally.

INITIATIVE—A GROWTH INGREDIENT

It is the consensus of those interested in professional development that the most critical period in young engineers' lives is the first few years after graduation from college. During this period, either they develop the healthy habit of growing on their own as professionals or they allow themselves to settle down into a routine of mediocre technical assignments.

During their undergraduate years in engineering school, they are under the tutelage, guidance, and direction of persons capable of training and guiding young people. It is here, in such an environment, that much can be done to mold the thinking in lines of professional development. When they leave school after commencement and join the ranks of industry or take their places in consultants' offices, they step into a brand new environment. In effect, they have crossed the bay, and the ocean stands before them.

The question of their future progress should be of vital interest and importance to all professional-minded young engineers. The measures they may take to ensure that progress depend not only upon their characters and environments, but also upon their points of view. They must inure themselves to the barbs of the unprofessional-minded, and be able to concentrate on the problem at hand without wavering in their determination.

Programming for Growth

It is a recognizable fact that the success of any venture depends upon developing people capable of carrying it out. This development takes time and much attendant planning for a program. The young graduate engineer must immediately set about to reappraise and reevaluate career objectives and ambitions and lay plans to meet and realize them. To do this properly and without lost motion, the help of an older, more experienced engineer is needed.

A program for growth does not allow, and has no place for, stifled initiative. If a company's professional program tends to base advancement on years of experience and practice rather than on individual ability, then the young professional's initiative becomes stifled. If such conditions are allowed to go undetected, they will become a matter of grave concern both to the professional-minded engineer and to management. For both, growth depends on the premiums that stimulate competitive effort.

The young engineer must recognize her or his unique relationship with management. The engineer's relationship to management is not one wherein a clear separation of interests exists. The young engineer's job is to develop the necessary ability to fill posts of greater responsibility if he or she wants to move ahead. Young professionals must do this in ways that will not prejudice their future. They must recognize this in projecting their thinking ahead and must govern themselves accordingly if they are to act effectively in enhancing their professional status.

Programming for growth holds no place for that erosion in thinking which appears to have been developing in recent years insofar as the engineer's professional attitude is concerned. It is quite evident that some of our younger engineers think too much about increased security and not enough about opportunity—too much of taking and not enough of giving.

This trend may reflect exposure to the "security thinking" so characteristic of our time. Although some younger engineers have caught the spark of enthusiasm and inspiration, many, lacking the inner drive, have become equipped with merely the "tools of the trade"—methods and techniques and basic information. They have become disinclined to make the effort to broaden and build on their technical and cultural foundation.

Programming for growth must include being able to rekindle the spirit of professionalism. "What security is there in it for me and how much more can I make?" must give way to "What opportunities lie ahead for me?" What is needed is an awakening of professional-mindedness as differentiated from merely technical-mindedness. All of us, as engineers, whether in the ranks or in management positions, should be interested in building a greater profession—not a better trade.

The young engineer must move either forward or backward; no one can remain static. If ambitions are energized, the movement is ahead. If a professional, young or old, allows herself or himself to fall prey to the anesthesia of easy security, he or she will not move ahead.

The Planned Program—Orientation into Industry

When young engineers leave the comparative safety of the college campus, they are practically incapable of analyzing problems or situations in terms of general principles. Until they can develop a degree of proficiency in this area, they tend to remain in the comparative isolation of technology. Industry at this point must take the raw recruits and adapt them to practical usefulness. The seed planted in a sound foundation by the colleges—the desire to obtain more knowledge and to develop professional competency—must here be nurtured.

A sound practical program of orientation to the workplace should provide the following:

1. An atmosphere favorable to professional development
2. On-the-job doing rather than merely observing
3. Evaluation and counseling on a continued basis
4. Progression toward greater responsibility
5. Establishment of an understanding that there is individual personal responsibility for professional self-development
6. Stimulants and incentives
7. Program assignments to utilize available talents fully
8. Adequate compensation and economic advancement

9. Encouragement to active participation in technical-society work
10. Some form of organizational ladder

As we look around us, we observe that there is considerable difference among companies in their consideration of what represents a professional attitude versus a subprofessional one. We hear some people say that management doesn't show as much interest as the colleges and universities in nurturing professional attitudes in young engineers. Perhaps these persons are merely being critical. There is much evidence today that the more progressive companies are doing an outstanding job of encouraging their engineers in the development of both technical competence and professional attitudes.

Probably the most important thing affecting the professional attitude of the young engineer in the first few years after graduation is the individual consideration received from the engineer's immediate boss and from his or her company. Of course, different companies handle this matter in different ways. The consulting engineering firms seem to have been making a much better impression on these younger people than many of the larger industries. There may be good reasons for this. Perhaps their necessarily individualistic approach in their business relations carries over to employees' work assignments and employee responsibility. But here again some of the larger industrial concerns, and smaller ones, too, are doing a good job in this area. Periodic individual rating, individual consideration of salary adjustments based on individual initiative and progress, and recognition of personal accomplishments all add to the necessary feeling of being wanted.

Now let us take a look at the things engineers, young or older, want from their employers:

1. Opportunity to do interesting, challenging, or important work and to have more freedom in or greater responsibility for one's work
2. Compensation commensurate with effort and ability
3. Desirable working conditions with respect to equipment, plant or office facilities, and the handling of service functions
4. Opportunity to work with competent and congenial coworkers
5. Opportunity for professional development, advancement, and recognition
6. Opportunity to work with competent supervisors
7. Control over matters of personal convenience and preference
8. Opportunity for advancement based on merit

The engineer finds satisfaction through personal advancement. Progressive management groups are alert to this, and they recognize that long-range profits depend on the steady movement *forward* of the most capable engineers. The best talent is put to work where it is needed most, and incentive is provided for top performance.

Let us now, for a moment, focus attention on the engineer, the personal requirements that the engineer must satisfy, and the responsibility the individual must shoulder. First, she or he must turn in a top performance. Employers look for it, and the engineer owes it to them. The engineer is being paid for this very thing.

Second, in order for an engineer to advance, he or she must prove to be the person best suited and qualified for promotion. Engineering is a creative profession and calls for creative talent, for the engineer must show how the present job can be done better than it has ever been done before.

Third, the young engineer must learn to help others do a better job, too. The qualities of good leadership go hand in hand with greater responsibility. The engineer who can demonstrate this kind of leadership will develop the essential characteristic that will necessarily lead to being given greater responsibility. It follows categorically that if young engineers are to move forward, they must accept responsibility for their own development, using all possible means to provide for continuing self-development.

As we look about us, we can see that the true professional's education is never ended. It can be put into use in bridging the gap between the engineering school and the industrial organization. Many industrial concerns have set up indoctrination programs as a means of introducing the graduate into their midst. This form of continuity or "gap-bridging" is all important. It quickly develops in the young engineer's mind a sense of stability, a sense of being wanted and accepted, a feeling of usefulness.

Although most cadet engineer training programs are scheduled for only one year, a continuation in some form for a much longer period is often desirable from both industrial and individual viewpoints. The essential requirement of industry, consultant's office, or field is a loyal, efficient, and aggressive working organization. Thus, the continued program will not only broaden technical competency but will also develop better human relationships early in the young engineer's career. Transfers from department to department in six-month or shorter intervals are advantageous. Concerns that do not have such flexibility can compensate for this shortcoming by giving the young engineer personal attention and greater opportunity for variety. Regimentation in training must be avoided if initiative is to be nurtured.

Employers have the power to inspire, guide, and assure financial assistance in the development of the professional character and complexion of the young engineer. Once a goal is established, the young engineer has climbed the first rung of the ladder to success. And this makes the difference between a professional-minded engineer and a technician. By putting responsibility into trainee jobs as early as possible, by making the trainee carry a burden and learn to produce, management can furnish the inspiration that will eliminate idle drifting. If management can set before these young people, and before older persons, too, recognized goals that are attainable, these goals will focus their attention almost automatically and develop them professionally.

Making the trainee a part of the working team is an essential part of any training or indoctrination program. To work, not merely to observe, should be the key to each training assignment. Since habits begin to be formed the first day on the job, the early formation of good habits constitutes a challenge. Work will bring out the best in anyone, and it will hold the really good employees.

CONTINUING EDUCATION

The educational atmosphere which the universities and colleges can offer is difficult to reproduce in the workplace. However, this atmosphere can contribute heavily to the professional development of young engineers when they are being oriented to industry, for it helps them realize that they should continue their education after college. Employers and education must bear joint responsibility for the advancement of the engineering profession as a whole. Engineering advances have wrought major changes in our concept of engineering education. How much knowledge of technical and scientific fundamentals one can absorb in a 4- or 5-year undergraduate college course is problematical. Thus, the workplace may be made to bear this after-commencement load and assume the greater portion of responsibility once the universities have indoctrinated the graduate engineer with the idea that a professional's education is never ended.

Active participation in technical-society work is one form that continuing education may take. When engineers contemplate the advantages of participation and membership in a professional society, they tend to consider with a true sense of value such aids to professional development as ready technical information, the rewarding associations and friendships that come from society contacts, and the stimulus to achievement that is born of the intercourse between keen, creative engineering minds. When they recognize these values, they may well ask, "What can I do for my society and the younger engineer?" They may become willing to lend a helping hand to aid in the general field of professional development.

The technical societies are the organizations with which engineering students have already been on "speaking terms." There is a good chance that they joined one of the student branches of a technical society and will continue membership after graduation. Young engineering graduates being oriented to industry or design office or groping for a happier outlet for their talents will often find that membership in a technical society gives them a certain prestige in the eyes of some of the more progressive employers. From direct contact at meetings or in committee work, they can get help and consideration from members of the society that will direct them into the channels most favorable to their talents. These are highly personal benefits that almost every engineer, at one time or another, desires and needs for professional development.

Employers can help foster the continuing education of their young engineers by encouraging company lectures and symposia geared to an academic level. Under this plan, the technical tools of the entire organization can be sharpened, and kept so. Company scholarships for advanced study are helpful in maintaining a high level of morale. Encouraging all engineering personnel to participate in evening courses in local approved institutions is another worthwhile effort and pays high dividends in the form of stability of organization and personnel.

Contribution in a highly technical area requires familiarity with the physical sciences far beyond the scope of regular undergraduate training. This requires that the young graduate so inclined consider continuing formal education on a full-time basis. The student should, however, keep in touch with the outside world in order to maintain a healthy balance between the human and scientific viewpoints. Tailoring individual progress to maintain this respect for and knowledge of people in perspective will add to a young engineer's professional stature. Scholarships and fellowship funds are available to the graduate for pursuing advanced education to some extent. The Graduate Studies Division of the American Society for Engineering Education (ASEE) is continually studying the situation in order to determine our present and future needs.

The ASEE reports that it has moved into a number of new areas of investigation, one of which is the project Evaluation of Engineering Education. The response and cooperation from faculty members in engineering colleges has been widespread. It has served to turn the attention of our engineering educators ahead to an evaluation of the needs of the engineering profession and the full-time educational needs of the engineer after graduation.

For the person who plans to be the most competent engineer in an organization, but who must of necessity work full-time, there are many avenues available for continuing study. In some instances a trip to the library will suffice; in others, there may be a need to return to school for one or more courses. Part-time degree programs at universities, noncredit courses, courses offered under state adult-education laws, and company-sponsored courses fall into this category of planned continuing education. Also, technical magazines provide much useful self-study information that can be a valuable asset to every engineer.

Continuing education as a prerequisite for reregistration is taking on added importance. At least one state (Iowa) requires continuing education as a prerequisite. Other states (for example Montana) require either continuing education or continued practice for reregistration. This practice will probably become more prevalent as time goes on.

If the engineer is dismayed at the prospect of returning to school, she or he may find it worthwhile to investigate correspondence courses. There are courses which give background in business administration and economics—essential to any management post—and others which can supply specific information and training in subjects ranging from accounting to writing. And since such courses will come in the mail, the engineer will not have to lose a minute from his or her regular job to take them. And now you can get continuing education courses from America's top engineering universities . . . right where you work! Further information on such courses, which are in the videocassette format, may be obtained by writing the Association for Media-Based Continuing Education for Engineers, Inc., 500 Tech Parkway, NW, Atlanta, Georgia 30313.

The professional should also develop skill in working with others effectively and in expressing ideas effectively. A long-range program should include preparation to undertake civic and social responsibilities. The professional should take interest in more broadening outlets of expression, such as popular lecture series and forums, which she or he should attend and participate in with increasing frequency. The scope of subjects will provide an intellectual challenge.

ETHICS—A PRIME ESSENTIAL

Professional standing is not merely the result of technical education. As we saw before, it is a matter of one's attitude and, in addition, of one's approach to work. Really, there is no direct connection between professional standing and advanced study of engineering technology. We can readily picture a doctor of engineering who is the world's authority on internal-combustion engines but who does not have a trace of professional understanding. On the other hand, there are a number of American engineers who appreciate the true meaning of professional status.

Earlier, we quoted Professor Wickenden as saying that what constitutes a professional is a matter of spirit and scope. We quote again from his book, *The Second Mile.*[16]

> To possess and to practice a special skill, even of high order, do not in themselves make an individual a professional man. Mere technical training at any level is vocational rather than professional. The difference between technical training and professional education is no simple matter of length—any difference of two years, or four, or six; nor is it a mere matter of intellectual difficulty. It is rather a matter of spirit and scope.

There you have it: spirit and scope. It should be apparent to all that professionals must be technically competent; they've got to know their business.

To explain still further this matter of spirit and scope, Vannevar Bush, in his well-known address "Qualities of a Profession" before the American Engineering Council in January, 1939, makes it quite clear:

> In every one of the professional groups, however, will be found the initial, central theme intact—they minister to the people. Otherwise they no longer endure as professional groups. Ministry carries with it the ideas of dignity and authority. It connotes no weakness and offers no apology. There is no fog of subservience surrounding the concept. The physician who ministers to his client takes charge by right of superior specialized knowledge of a highly personal aspect of the affairs of the individual. The attorney assumes professional responsibility for guiding the legal acts of his client and speaks with the whole authority of the statutes as a background. It is in this higher sense that we trace the thread of ministry to the people.

What Bush has said checks closely with the opinions of philosophers, scholars, and the general public. Thus, we can safely conclude that the principal requirement for professional standing is to minister to the public.

Now the professional who ministers to the people must be trustworthy and morally responsible. In short, ethical.

The layperson looks to the professional because the former knows nothing about the latter's technology. The layperson has to depend entirely on the professional's integrity. The people have to take the engineer's word for it that there has been no cheating to gain a dollar here or there when the engineer:

Purifies city drinking water

Insulates electric wires and apparatus

Specifies materials for skyscrapers

Designs linkages for airplane controls

Exhausts dangerous fumes from buildings

Designs the intricate controls for automatic elevators

The public has to rely upon the integrity of engineers when they design and manufacture automobiles or supersonic planes. But this reliance is misplaced if the engineer is not ethical.

Engineers, young or older, should learn more about ethics and do some honest worrying about their application in everyday living and working. They should prevail upon the officers of the local engineering society to organize meetings for the discussion of ethics. They should prevail upon the officers of the national engineering society to include sessions on ethics regularly in national and regional programs. Before they are hired, engineers should be asked what they know about ethics. Editors of their favorite magazines should be requested to publish articles on this topic. All should make it a practice to bring up the topic when gathered in groups. The older engineer should impress the younger engineer with the practical manifestations of ethics.

For your convenience, the Code of Ethics of Engineers, prepared by ABET, is reproduced in Fig. 1.1. Read it carefully, even if you have read it before. The Code of Ethics for Engineers, published by the National Society of Professional Engineers (NSPE) in 1985, is in Fig. 1.2.

The National Council of Engineering Examiners (NCEE) Committee of Professionalism and Ethics is active in revising the Model Rules of Professional Conduct. Many state boards have adopted a code of ethics. In fact, some boards give a take-home exam on state registration laws and their adopted code of ethics or rules of professional conduct.

RULES OF PROFESSIONAL CONDUCT

State boards of registration have taken on the added responsibility of policing the engineering profession for both legal and ethical transgressions of a serious nature. They have developed guidelines and rules of procedure that are meaningful, and they have provided for their enforcement. Substantive guidelines have been well developed and spelled out by NCEE[21] in the form of suggested rules of professional conduct which are based on those parts of the engineering code of ethics that directly

*Accreditation Board for Engineering and Technology**

CODE OF ETHICS OF ENGINEERS

THE FUNDAMENTAL PRINCIPLES

Engineers uphold and advance the integrity, honor and dignity of the engineering profession by:

I. using their knowledge and skill for the enhancement of human welfare;

II. being honest and impartial, and serving with fidelity the public, their employers and clients;

III. striving to increase the competence and prestige of the engineering profession; and

IV. supporting the professional and technical societies of their disciplines.

THE FUNDAMENTAL CANONS

1. Engineers shall hold paramount the safety, health and welfare of the public in the performance of their professional duties.

2. Engineers shall perform services only in the areas of their competence.

3. Engineers shall issue public statements only in an objective and truthful manner.

4. Engineers shall act in professional matters for each employer or client as faithful agents or trustees, and shall avoid conflicts of interest.

5. Engineers shall build their professional reputation on the merit of their services and shall not compete unfairly with others.

6. Engineers shall act in such a manner as to uphold and enhance the honor, integrity and dignity of the profession.

7. Engineers shall continue their professional development throughout their careers and shall provide opportunities for the professional development of those engineers under their supervision.

345 East 47th Street New York, NY 10017

*Formerly Engineers' Council for Professional Development. (Approved by the ECPD Board of Directors, October 5, 1977)

AB-54 2/85

Figure 1.1.

relate to protecting the public health, safety, and welfare.

To remove any doubt about the authority of state boards, many state laws have been amended to prescribe specifically the right and duty of the boards to adopt rules of professional conduct and to spell out clearly that the violation of such rules may be grounds for suspension or revocation of the license of the individual involved. Their enforcement is effective.

In almost every state, there are discussions going on about what changes state societies or their members might like to see enacted. The Texas Society of Professional Engineers (TSPE) has put through an amendment to give the Texas State Board of Registration the right to increase the annual renewal fees so that the board will have the funds to implement a vigorous enforcement campaign. We can expect this movement to spread to other states.

In some states, efforts are being made to eliminate or restrict exemptions from licensure for those holding public office in an engineering capacity. In still other states, work is under way to provide better definition of the meaning of "responsible charge" of engineering work in the day-to-day performance of engineering activities in the design office and field. We can look at each state, therefore, as a laboratory, and from its experimentation and experiences the engineering profession will be able to analyze, debate, and act upon the constant wave of revisions to the state laws, which are intended to better serve the public. See Fig. 1.3, "Model Rules of Professional Conduct, National Council of Engineering Examiners (1985)."

CODE OF ETHICS FOR ENGINEERS

PREAMBLE

Engineering is an important and learned profession. The members of the profession recognize that their work has a direct and vital impact on the quality of life for all people. Accordingly, the services provided by engineers require honesty, impartiality, fairness and equity, and must be dedicated to the protection of the public health, safety and welfare. In the practice of their profession, engineers must perform under a standard of professional behavior which requires adherence to the highest principles of ethical conduct on behalf of the public, clients, employers and the profession.

I. FUNDAMENTAL CANONS

Engineers, in the fulfillment of their professional duties, shall:

1. Hold paramount the safety, health and welfare of the public in the performance of their professional duties.
2. Perform services only in areas of their competence.
3. Issue public statements only in an objective and truthful manner.

Figure 1.2.

4. Act in professional matters for each employer or client as faithful agents or trustees.
5. Avoid improper solicitation of professional employment.

II. RULES OF PRACTICE

1. Engineers shall hold paramount the safety, health and welfare of the public in the performance of their professional duties.
   a. Engineers shall at all times recognize that their primary obligation is to protect the safety, health, property and welfare of the public. If their professional judgment is overruled under circumstances where the safety, health, property or welfare of the public are endangered, they shall notify their employer or client and such other authority as may be appropriate.
   b. Engineers shall approve only those engineering documents which are safe for public health, property and welfare in conformity with accepted standards.
   c. Engineers shall not reveal facts, data or information obtained in a professional capacity without the prior consent of the client or employer except as authorized or required by law or this Code.
   d. Engineers shall not permit the use of their name or firm name nor associate in business ventures with any person or firm which they have reason to believe is engaging in fraudulent or dishonest business or professional practices.
   e. Engineers having knowledge of any alleged violation of this Code shall cooperate with the proper authorities in furnishing such information or assistance as may be required.
2. Engineers shall perform services only in the areas of their competence:
   a. Engineers shall undertake assignments only when qualified by education or experience in the specific technical fields involved.
   b. Engineers shall not affix their signatures to any plans or documents dealing with subject matter in which they lack competence, nor to any plan or document not prepared under their direction and control.
   c. Engineers may accept assignments and assume responsibility for coordination of an entire project and sign and seal the engineering documents for the entire project, provided that each technical segment is signed and sealed only by the qualified engineers who prepared the segment.
3. Engineers shall issue statements only in an objective and truthful manner.
   a. Engineers shall be objective and truthful in professional reports, statements or testimony. They shall include all relevant and pertinent information in such reports, statements or testimony.
   b. Engineers may express publicly a professional opinion on technical subjects only when that opinion is founded upon adequate knowledge of the facts and competence in the subject matter.
   c. Engineers shall issue no statements, criticisms or arguments on technical matters which are inspired or paid for by interested parties, unless they have prefaced their comments by explicitly identifying the interested parties on whose behalf they are speaking, and by

Figure 1.2. *(Continued)*

revealing the existence of any interest the engineers may have in the matters.

4. Engineers shall act in professional matters for each employer or client as faithful agents or trustees.
   a. Engineers shall disclose all known or potential conflicts of interest to their employers or clients by promptly informing them of any business association, interest, or other circumstances which could influence or appear to influence their judgment or the quality of their services.
   b. Engineers shall not accept compensation, financial or otherwise, from more than one party for services on the same project, or for services pertaining to the same project, unless the circumstances are fully disclosed to, and agreed to by, all interested parties.
   c. Engineers shall not solicit or accept financial or other valuable consideration, directly or indirectly, from contractors, their agents, or other parties in connection with work for employers or clients for which they are responsible.
   d. Engineers in public service as members, advisors or employees of a governmental body or department shall not participate in decisions with respect to professional services solicited or provided by them or their organizations in private or public engineering practice.
   e. Engineers shall not solicit or accept a professional contract from a governmental body on which a principal or officer of their organization serves as a member.
5. Engineers shall avoid improper solicitation of professional employment.
   a. Engineers shall not falsify or permit misrepresentation of their, or their associates', academic or professional qualifications. They shall not misrepresent or exaggerate their degree of responsibility in or for the subject matter of prior assignments. Brochures or other presentations incident to the solicitation of employment shall not misrepresent pertinent facts concerning employers, employees, associates, joint ventures or past accomplishments with the intent and purpose of enhancing their qualifications and their work.
   b. Engineers shall not offer, give, solicit or receive, either directly or indirectly, any political contribution in an amount intended to influence the award of a contract by public authority, or which may be reasonably construed by the public of having the effect or intent to influence the award of a contract. They shall not offer any gift, or other valuable consideration in order to secure work. They shall not pay a commission, percentage or brokerage fee in order to secure work except to a bona fide employee or bona fide established commercial or marketing agencies retained by them.

III. PROFESSIONAL OBLIGATIONS

1. Engineers shall be guided in all their professional relations by the highest standards of integrity.
   a. Engineers shall admit and accept their own errors when proven wrong and refrain from distorting or altering the facts in an attempt to justify their decisions.

Figure 1.2. *(Continued)*

b. Engineers shall devise their clients or employers when they believe a project will not be successful.
c. Engineers shall not accept outside employment to the detriment of their regular work or interest. Before accepting any outside employment, they will notify their employers.
d. Engineers shall not attempt to attract an engineer from another employer by false or misleading pretenses.
e. Engineers shall not actively participate in strikes, picket lines, or other collective coercive action.
f. Engineers shall avoid any act tending to promote their own interest at the expense of the dignity and integrity of the profession.

2. Engineers shall at all times strive to serve the public interest.
   a. Engineers shall seek opportunities to be of constructive service in civic affairs and work for the advancement of the safety, health and well-being of their community.
   b. Engineers shall not complete, sign, or seal plans and/or specifications that are not of a design safe to the public health and welfare and in conformity with accepted engineering standards. If the client or employer insists on such unprofessional conduct, they shall notify the proper authorities and withdraw from further service on the project.
   c. Engineers shall endeavor to extend public knowledge and appreciation of engineering and its achievements and to protect the engineering profession from misrepresentation and misunderstanding.
3. Engineers shall avoid all conduct or practice which is likely to discredit the profession or deceive the public.
   a. Engineers shall avoid the use of statements containing a material misrepresentation of fact or omitting a material fact necessary to keep statements from being misleading; statements intended or likely to create an unjustified expectation; statements containing prediction of future success; statements containing an opinion as to the quality of the Engineers' services; or statements intended or likely to attract clients by the use of showmanship, puffery, or self-laudation, including the use of slogans, jingles, or sensational language or format.
   b. Consistent with the foregoing, Engineers may advertise for recruitment of personnel.
   c. Consistent with the foregoing, Engineers may prepare articles for the lay or tehcnical press, but such articles shall not imply credit to the author for work performed by others.
4. Engineers shall not disclose confidential information concerning the business affairs or technical processes of any present or former client or employer without his consent.
   a. Engineers in the employ of others shall not without the consent of all interested parties enter promotional efforts or negotiations for work or make arrangements for other employment as a principal or to practice in connection with a specific project for which the Engineer has gained particular and specialized knowledge.
   b. Engineers shall not, without the consent of all interested parties,

Figure 1.2. (*Continued*)

participate in or represent an adversary interest in connection with a specific project or proceeding in which the Engineer has gained particular specialized knowledge on behalf of a former client or employer.

5. Engineers shall not be influenced in their professional duties by conflicting interests.
   a. Engineers shall not accept financial or other considerations, including free engineering designs, from material or equipment suppliers for specifying their product.
   b. Engineers shall not accept commissions or allowances, directly or indirectly, from contractors or other parties dealing with clients or employers of the Engineer in connection with work for which the Engineer is responsible.
6. Engineers shall uphold the principle of appropriate and adequate compensation for those engaged in engineering work.
   a. Engineers shall not accept remuneration from either an employee or employment agency for giving employment.
   b. Engineers, when employing other engineers, shall offer a salary according to professional qualifications and the recognized standards in the particular geographical area.
7. Engineers shall not compete unfairly with other engineers by attempting to obtain employment or advancement or professional engagements by taking advantage of a salaried position, by criticizing other engineers, or by other improper or questionable methods.
   a. Engineers shall not request, propose, or accept a professional commission on a contingent basis under circumstances in which their professional judgment may be compromised.
   b. Engineers in salaried positions shall accept part-time engineering work only at salaries not less than that recognized as standard in the area.
   c. Engineers shall not use equipment, supplies, laboratory or office facilities of an employer to carry on outside private practice without consent.
8. Engineers shall not attempt to injure, maliciously or falsely, directly or indirectly, the professional reputation, prospects, practice or employment of other engineers, nor indiscriminately criticize other engineers' work. Engineers who believe others are guilty of unethical or illegal practice shall present such information to the proper authority for action.
   a. Engineers in private practice shall not review the work of another engineer for the same client, except with the knowledge of such engineer, or unless the connection of such engineer with the work has been terminated.
   b. Engineers in governmental, industrial or educational employ are entitled to review and evaluate the work of other engineers when so required by their employment duties.
   c. Engineers in sales or industrial employ are entitled to make engineering comparisons of represented products with products of other suppliers.

Figure 1.2. (*Continued*)

9. Engineers shall accept personal responsibility for all professional activities.
   a. Engineers shall conform with state registration laws in the practice of engineering.
   b. Engineers shall not use association with a nonengineer, a corporation, or partnership as a "cloak" for unethical acts, but must accept personal responsibility for all professional acts.
10. Engineers shall give credit for engineering work to those to whom credit is due, and will recognize the proprietary interests of others.
    a. Engineers shall, whenever possible, name the person or persons who may be individually responsible for designs, inventions, writings or other accomplishments.
    b. Engineers using designs supplied by a client recognize that the designs remain the property of the client and may not be duplicated by the Engineer for others without express permission.
    c. Engineers, before undertaking work for others in connection with which the Engineer may make imrpovements, plans, designs, inventions or other records which may justify copyrights or patents, should enter into a positive agreement regarding ownership.
    d. Engineers' designs, data, records and notes referring exclusively to an employer's work are the employer's property.
11. Engineers shall cooperate in extending the effectiveness of the profession by interchanging information and experience with other engineers and students, and will endeavor to provide opportunity for the professional development and advancement of engineers under their supervision.
    a. Engineers shall encourage engineering employees' efforts to improve their education.
    b. Engineers shall encourage engineering employees to attend and present papers at professional technical society meetings.
    c. Engineers shall urge engineering employees to become registered at the earliest possible date.
    d. Engineers shall assign a professional engineer duties of a nature to utilize full training and experience, insofar as possible, and delegate lesser functions to subprofessionals or to technicians.
    e. Engineers shall provide a prospective engineering employee with complete information on working conditions and proposed status of employment, and after employment will keep employees informed of any changes.

"By order of the United States District Court for the District of Columbia, former Section11(c) of the NSPE Code of Ethics prohibiting competitive bidding, and all policy statements, opinions, rulings or other guidelines interpreting its scope, have been rescinded as unlawfully interfering with the legal right of engineers, protected under the antitrust laws, to provide price information to prospective clients; accordingly, nothing contained in the NSPE Code of Ethics, policy statements, opinions, rulings or other guidelines prohibits the submission of price quotations or competitive bids for engineering services at any time or in any amount."

Figure 1.2. *(Continued)*

Statement by NSPE Executive Committee

In order to correct misunderstandings which have been indicated in some instances since the issuance of the Supreme Court decision and the entry of the Final Judgment, it is noted that in its decision of April 25, 1978, the Supreme Court of the United States declared: "The Sherman Act does not require competitive bidding."

It is further noted that as made clear in the Supreme Court decision:

1. Engineers and firms may individually refuse to bid for engineering services.
2. Clients are not required to seek bids for engineering services.
3. Federal, state, and local laws governing procedures to procure engineering services are not affected, and remain in full force and effect.
4. State societies and local chapters are free to actively and aggressively seek legislation for professional selection and negotiation procedures by public agencies.
5. State registration board rules of professional conduct, including rules prohibiting competitive bidding for engineering services, are not affected and remain in full force and effect. State registration boards with authority to adopt rules of professional conduct may adopt rules governing procedures to obtain engineering services.
6. As noted by the Supreme Court, "nothing in the judgement prevents NSPE and its members from attempting to influence governmental action. . . ."

NOTE: In regard to the question of application of the Code to corporations vis-a-vis real persons, business form or type should not negate nor influence conformance of individuals to the Code. The Code deals with professional services, which services must be performed by real persons. Real persons in turn establish and implement policies within business structures. The Code is clearly written to apply to the Engineer and it is incumbent on a member of NSPE to endeavor to live up to its provisions. This applies to all pertinent sections of the Code.

NSPE Publication No. 1102 as revised January 1985

Figure 1.2. *(Continued)*

Character—Chief Element of Success

One of the most important stepping-stones to the successful development of a professional career is character. Character is one of those words which is understood by all, but difficult to define. These two definitions, however, are worthy of quoting: "Character is the sum of the inherited and acquired ethical traits which give a person his moral individuality," and "A character, or that which distinguishes one man from all others, cannot be supposed to consist of one particular virtue or vice or passion only, but it is a composition of qualities which are not contrary to one another in the same person."

Whenever we think of an important undertaking that has been successfully carried through, any big enterprise that has been successfully managed, or any big company that has justified its existence, we immediately associate with these ventures those individuals who were responsible for their success. The imprint of strong character on these organizations is so enduring that for a considerable time it continues to glow and carries on the reputation of the people responsible.

Engineers, to be truly and successfully professional, should develop confidence in their relations with their fellow engineers and business associates. We have all been associated, at one time or another, with persons in whom we had perfect confidence, with some in whom we had no confidence at all, and with others in whom we were undecided whether or not to place confidence. What a pleasure it is to work with someone in complete confidence, with no tension, and a minimum of nervous energy required. This is the relationship which forms the foundation for a successful professional life. Progress is not made without confidence, and without character, confidence is nonexistent. Without character there is no permanency. "Caissons are more important than towers," so the saying goes. Getting down to bedrock to build a foundation for a skyscraper is necessary for the structure to endure.

Some engineers educated in our colleges seem to have become so preoccupied with getting on in life that they have failed to realize that a good understanding of the intrinsic worth of virtue is far more valuable and far more important than a mere acquisition of knowledge that misses this understanding. Character-building should be a required course in every year of a person's life, and everyone should be shown how to apply it to furthering individual and national prosperity.

The character trait of desiring self-improvement should be nurtured while young people are still in school. They must not be allowed to leave school with the thought that their diplomas mark the end of their journey in education; the end of formal schooling must be considered really and truly "commencement." The sheepskin is only the foundation upon which they must continue to build and they themselves must do the building. This point cannot be overemphasized.

Optimism and cheerfulness should be cultivated, for they have a definite place in the complexion of a successful professional career. We can all work through the pleasant duties when the going is easy. But cheerfulness and optimism carry us over the difficult spots, the hard places. This attitude makes the difference between success and failure. Such an outlook will solve even the most difficult problems in professional everyday life.

The engineer should look up from daily preoccupations and develop a reading acquaintance with good authors. ABET reports that a good reading program should include three essential parts:

1. General readings in biography, travel, history, economics, sociology, psychology, philosophy, natural science, and literature, including

NATIONAL COUNCIL OF ENGINEERING EXAMINERS

MODEL RULES OF PROFESSIONAL CONDUCT

A GUIDE FOR USE BY REGISTRATION BOARDS

Figure 1.3.

NCEE MODEL RULES OF PROFESSIONAL CONDUCT
August 1985

Preamble

To comply with the purpose of the (identify State Registration Statute) which is to safeguard life, health and property, to promote the public welfare, and to maintain a high standard of integrity and practice, the (identify State Board Registration Statute) has developed the following Rules of Professional Conduct. These rules shall be developed binding on every person holding a certificate of registration to offer or perform engineering or land surveying services in this state. All persons registered under (identify State Registration Statute) are required to be familiar with the Registration Statute and these rules. The Rules of Professional Conduct delineate specific obligations the registrant must meet. In addition, each registrant is charged with the responsibility of adhering to standards of highest ethical and moral conduct in all aspects of the practice of Professional Engineering and Land Surveying.

The practice of Professional Engineering and Land Surveying is a privilege, as opposed to a right. All registrants shall exercise their privilege of practicing by performing services only in the areas of their competence according to current standards of technical competence.

Registrants shall recognize their responsibility to the public and shall represent themselves before the public only in an objective and truthful manner.

They shall avoid conflicts of interest and faithfully serve the legitimate interests of their employers, clients, and customers within the limits defined by these rules. Their professional reputation shall be built on the merit of their services and they shall not compete unfairly with others.

The Rules of Professional Conduct as promulgated herein are enforced under the powers vested by (identify State Enforcing Agency). In these Rules, the word "registrant" shall mean any person holding a license or a certificate issued by (identify State Registration Agency).

RULES OF PROFESSIONAL CONDUCT

I. Registrant's Obligation to Society

a. Registrants, in the performance of their services for clients, employers and customers, shall be cognizant that their first and foremost responsibilty is to the public welfare.
b. Registrants shall approve and seal only those design documents and surveys that conform to accepted engineering and land surveying standards and safeguard the life, health, property and welfare of the public.
c. Registrants shall notify their employer or client and such other authority as may be appropriate when their professional judgment is overruled under circumstances where the life, health, property or welfare of the public is endangered.
d. Registrants shall be objective and truthful in professional reports, statements or testimony. They shall include all relevant and pertinent information in such reports, statements or testimony.

Figure 1.3. (*Continued*)

e. Registrants shall express a professional opinion publicly only when it is founded upon an adequate knowledge of the facts and a competent evaluation of the subject matter.

f. Registrants shall issue no statements, criticisms or arguments on technical matters which are inspired or paid for by interested parties, unless they explicitly identify the interested parties on whose behalf they are speaking, and reveal any interest they have in the matters.

g. Registrants shall not permit the use of their name or firm name by, nor associate in business ventures with, any person or firm which is engaging in fraudulent or dishonest business or professional practices.

h. Registrants having knowledge of possible violations of any of these RULES OF PROFESSIONAL CONDUCT shall provide the State Board information and assistance necessary to the final determination of such violation.

II. Registrant's Obligation to Employer and Clients

a. Registrants shall undertake assignments only when qualified by education or experience in the specific technical fields of engineering or land surveying involved.

b. Registrants shall not affix their signatures or seals to any plans or documents dealing with subject matter in which they lack competence, nor to any such plan or document not prepared under their direct control and personal supervision.

c. Registrants may accept assignments for coordination of an entire project, provided that each design segment is signed and sealed by the registrant responsible for preparation of that design segment.

d. Registrants shall not reveal facts, data or information obtained in a professional capacity without the prior consent of the client or employer except as authorized or required by law.

e. Registrants shall not solicit or accept financial or other valuable consideration, directly or indirectly, from contractors, their agents or other parties in connection with work for employers or clients.

f. Registrants shall make full prior disclosures to their employers or clients of potential conflicts of interest or other circumstances which could influence or appear to influence their judgment or the quality of their service.

g. Registrants shall not accept compensation, financial or otherwise, from more than one party for services pertaining to the same project, unless the circumstances are fully disclosed and agreed to by all interested parties.

h. Registrants shall not solicit or accept a professional contract from a governmental body on which a principal or officer of their organization serves as a member. Conversely, registrants serving as members, advisors, or employees of a governmental body or department, who are the principals or employees of a private concern, shall not participate in decisions with respect to professional services offered or provided by said concern to the governmental body which they serve.

III. Registrant's Obligation to Other Registrants

a. Registrants shall not falsify or permit misrepresentation of their, or their

Figure 1.3. *(Continued)*

associates', academic or professional qualifications. They shall not misrepresent or exaggerate their degree of responsibility in prior assignments nor the complexity of said assignments. Presentations incident to the solicitation of employment or business shall not misrepresent pertinent facts concerning employers, employees, associates, joint ventures or past accomplishments.

b. Registrants shall not offer, give, solicit or receive, either directly or indirectly, any commission, or gift, or other valuable consideration in order to secure work, and shall not make any political contribution with the intent to influence the award of a contract by public authority.
c. Registrants shall not attempt to injure, maliciously or falsely, directly or indirectly, the professional reputation, prospects, practice or employment of other registrants, nor indiscriminately criticize other registrants' work.

Figure 1.3. (*Continued*)

fiction and essays represented in the ABET Reading List for Engineers.

2. Technical readings, as represented by books listed in the various sections of the ABET Selected Bibliography of Engineering Subjects.
3. Professional reading, such as that made available by the periodicals of the national societies and industry.

Professional Identification

The many benefits of professional identification for the young engineer at graduation suggest becoming actively associated with an engineering group and qualifying for the preliminary step to professional registration through the Engineer-in-Training (EIT) program. There has been wide acceptance of the movement, started in 1943, to interest the young engineer in further professional development by offering at the time of graduation from college the fundamental and engineering-science parts of the written examination required for professional licensure. State boards now offer the FE examination to engineering seniors and in some cases, juniors as well.

Generally, these examinations are conducted by boards on the campuses of engineering schools with accredited ABET programs. Candidates who successfully pass this examination are given an EIT certificate, which will exempt them from the first day of the two-day written examination in the state in which it is issued when they apply for registration as professional engineers.

Since all states provide the same examination, they may recognize the original EIT certificate. Since most engineering college students will be required to pass a two-day written examination as one of the requirements for registration unless they have the EIT certificate, and since many ambitious young engineers aspire to positions of responsibility (which will most likely require registration), many engineering college students will be taking the FE examination before graduation.

Engineers' Registration is the legal manifestation of a professional career. Every engineer should be familiar with the laws regulating the practice of professional engineering in his or her home state. Many states examine an applicant on their state law even if the applicant is licensed elsewhere.

ENGINEERING COMPETENCE

To many proponents of mandatory requalification for reregistration, particularly the ill-informed lay consumer interests, competence is supposed to be some simply defined attribute of a professional to be readily measured by boards of registration, and which, if the board determines that an individual has it, permits assurance to the public that that individual can solve any problem effectively so long as it is engineering in nature.

Nothing can be further from the truth. *Competence*, defined as the ability to solve engineering problems, is a highly individual characteristic. It varies in both nature and degree from one individual to another with infinite variety. Each engineer has a different ability to solve the spectrum of various problems which may confront her or him in the engineering team. Each individual can successfully solve only a fraction of those problems he or she meets, depending upon the fund of engineering knowledge acquired by education and training on the job, native abilities, judgment, and so on. Some specialists can perform brilliantly within a narrow range of problems, others only moderately over a broader scope.

The individual engineer is the best judge of his or her own *total competence*. Therefore, most registration laws have at least implicitly placed the burden of proof of competence on the registrant as an ethical requirement. Most boards of registration have the authority to revoke a license on evidence of breach of ethical conduct in this and other respects.

What then is the nature of the "competence" presently recognized in our registration process as a basis for issuing licenses to enter the profession of engineering? How can this be demonstrated?

Long experience has demonstrated that this process results in a group of professionals who with a very few exceptions can deal successfully with engineering problems which they choose to attempt. This process is (1) board-accredited engineering education, (2) a certifying FE examination, (3) hands-on experience or training in the engineering community in the company of practicing professionals, usually for a minimum of four years, (4) a second examination intended to measure in a limited degree the ability to apply fundamentals to practical problems, and finally (5) accreditation by professional peer references regarding early performance and ethical attitudes.

2

WHY ENGINEERS' REGISTRATION?

Legal registration of engineers has gone far beyond a theory—it is an established fact that is here to stay. Because the pioneers of the registration movement had the "vision and strong belief and conviction that it represented an indispensable forward step of progress in the profession, because they overcame complacency and disparagement, prejudice and misunderstanding, secret opposition and open antagonism," Engineers' Registration has become law in every state and territorial possession that has enacted statutes governing the practice of engineering by individuals.

HISTORY

The progress of enactment of engineers' registration laws dates from Wyoming in 1907. As our population became more dense, the accumulation and increasing number of wrong guesses in construction fields served to focus the attention of the American public upon the engineering profession and the need for an authoritative regulation of its practitioners. The fateful day came in 1907 in Wyoming.

Wyoming had embarked upon a vast irrigation program. People flocked to the areas affected from all parts of the United States. Then, since engineering often requires little capital investment, many of these people sought to do the engineering portion of the work. The situation which developed was bad, and more bad guesses were imminent. Impossible things were proposed, and, we are told, some self-styled engineers planned systems which would have required water to flow uphill without the help of pumping equipment.

To meet the emergency, State Engineer Clarence T. Johnston went to the legislature and insisted that "for the protection of the people of the State of Wyoming, we must regulate by law the practice of engineering." He was told to prepare a bill and was assured that it would be enacted. Thus, in 1907, Wyoming became the first state to have an engineering law.

In the years that followed, progress was slow. By 1919 only four states had enacted engineers' registration laws. The real battle covered the last thirty years of the full forty-year concentrated program to regulate the profession. By 1921, the number of states increased to twenty. Then, for a time, real progress seemed halted, with only eight more states gained during the following eleven years and no further gains in sight. Forces of opposition had gathered strength and were blocking real progress, and it even seemed that they were going to reverse the tide. The entire registration movement seemed to be hanging in the balance.

Then, in 1934, the deadlock was broken. The friends of Engineers' Registration gathered together their forces, now national in scope, and, adding impetus where needed most, turned almost sure defeat into victory. The following year seven more states were added to the roster. Not until 1947 did the last state, Montana, enact engineering statutes. This completed the roster for the states. The District of Columbia acquired its law in 1950, so by then we had engineering laws in the fifty states (Hawaii and Alaska enacted laws in 1923 and 1939 respectively, before their statehood), the District of Columbia, and Puerto Rico. Table 2.1 is a listing of NCEE member boards.

In 1920, there were only a handful or registered professional engineers in the United States. By 1930, 10,000 engineers possessed legal status. In 1986 there were 343,000 registered professional engineers, with this number increasing rapidly each year. As licensing procedures are simplified and strengthened and more engineers are made more aware of the long-range benefits, the practice of seeking and obtaining licensure will become more widespread.

LEGAL BASIS FOR REGISTRATION

To a state legislature, engineering registration is a means of protecting the life, health, and property of the public. Legal registration permits only those engineers who can safely serve the public to practice and denies such responsibility to those who are proved incompetent. When minimum standards for registration are prescribed, incompetents are

TABLE 2.1 Member Board of NCEE by Zones (Date Year Law Enacted)

		Western Zone			
Alaska	1939	Hawaii	1923	Oregon	1919
Arizona	1921	Idaho	1919	Utah	1935
California	1929	Montana	1947	Washington	1935
Colorado	1919	Nevada	1919	Wyoming	1907
Guam	1960	New Mexico	1935	Northern Mariana Is.	1978
		Central Zone			
Illinois	1945	Michigan	1919	North Dakota	1943
Indiana	1921	Minnesota	1921	Ohio	1933
Iowa	1919	Missouri	1941	South Dakota	1925
Kansas	1931	Nebraska	1937	Wisconsin	1931
Kentucky	1938				
		Southern Zone			
Alabama	1935	Mississippi	1928	Tennessee	1921
Arkansas	1925	North Carolina	1921	Texas	1937
Florida	1917	Oklahoma	1935	Virgin Is. (U.S)	1968
Georgia	1937	Puerto Rico	1927		
Louisiana	1908	South Carolina	1922		
		Northeast Zone			
Connecticut	1935	Massachusetts	1941	Rhode Island	1938
Delaware	1941	New Hampshire	1945	Vermont	1939
District of Columbia	1950	New Jersey	1921	Virginia	1920
Maine	1935	New York	1920	West Virginia	1921
Maryland	1939	Pennsylvania	1921		

NOTE: Above dates apply solely to the PE Boards and do not include the LS Boards. Table courtesy NCEE (1985).

prevented from practicing engineering, i.e., from representing themselves to the public as licensed professional engineers.

Legal registration obtains in other professions, such as medicine, architecture, dentistry, and nursing. All the professions, except the ministry, are regulated under the police powers of the state. Government has the right to impose sufficient restraint on the actions of its citizens to conserve the enjoyment of the rights of the individual. The power of the government to impose this restraint is called the police power, and it is broad enough to protect the citizens of the state in the exercise of their legal or natural rights from injury by others.

The state has a right to determine who shall practice any occupation where the protection of property and the safety, welfare, and health of the general public are involved. The police power of the state is used, therefore, to protect the public from quacks and incompetents. The public needs to be protected against the unscrupulous and the imposter, who do not belong in a profession but nevertheless practice in its name. A doctor cannot prescribe medicine or perform an operation unless she or he is professionally registered, a lawyer cannot draw up a will for someone else unless he or she has passed the bar examination.

It wasn't always this way. Once upon a time, kings appointed physicians; and, not too long ago, a judge could examine an applicant in open court and admit that person to the bar.

In what era this police power we have been talking about first made felt its restraining action for the protection of the public health and safety has never been recorded and is not known, but it must date far back in the history of government. In engineering, the problem of such protection was recognized in one form or other as far back as 2000 B.C. Then, a code of laws said:

> If a builder erect a house for a man and do not make its construction firm, and the house which he built collapse and cause the death of the owner of the house, that builder shall be put to death.
>
> If it cause the death of the son of the owner of the house, they shall put to death the son of the builder.

Thus read sections 229 and 230 of the Laws of Hammurabi of Babylon, almost 4000 years ago. No provision was made for registering engineers or contractors, but we see that ample provision certainly was made to punish the incompetent and the imposter.

In more modern times, another example of the exercise of police power on the state made itself manifest in the Chicago fire. Police assumed the right to dynamite undamaged buildings to prevent the spread of the fire to other and more distant buildings. Should an epidemic break out, the state has the right to quarantine the infected area to prevent spread of the disease. And it follows that the state has the right to prevent a hazardous situation by evaluating and passing on the qualifications of those practitioners whose activities have obvious public implications: the physician, the lawyer, the engineer. The elimination and exclusion, as a matter of public welfare, of the dishonest and unqualified from the practice of these professions is the state's undebatable right.

A profession is a calling or vocation requiring mental rather than manual labor. Licensure is certification of qualifications in advance of practice. Usually, the statute regulating this practice will contain certain general requirements, but always the administration of these requirements is left to the profession itself, i.e., to the state boards.

The public reserves a particular field of endeavor for the exclusive practice of those who have been legally recognized as qualified in that field. It gives the practitioner a special place in the courts and protects the professional from the unqualified opinions of laypersons. In a measure, a professional practitioner possesses a special form of citizenship. In return, the public expects many things; but, in particular, it wants competence and protection from bungling and extortion.

Regulation of Engineers' Registration, implicit in every state because of its police powers, is achieved in one of two ways, either by protecting the use of the title "professional engineer" or by regulating the practice of the profession.* Both methods have been declared constitutional by the courts. Most civil courts have adopted the general policy that expert engineering testimony may not be received from unregistered individuals. They have further taken the stand that an unregistered engineer cannot recover fees for engineering services, and such cases are often dismissed without trial. In one instance, an apparently registered engineer was not able to collect because he was not in paid-up good standing with his state registration office insofar as his registration fee was concerned.

In New York State, a recent case involved a mail-order house which was advertising free engineering services in connection with the sale of heating equipment. The matter did not reach the courts, since the company agreed to discontinue such advertising when the legal situation was brought to its attention. In this connection, the new provision in the law of New York State for procedure by injunction against violators proved to be a most effective deterrent.

Although the various states strive for unity in putting their registration laws to work, uniformity is difficult, because each state has rights as an independent commonwealth and the prerogative to deviate in accord-

---

* In a number of states, the title "engineer" is protected as well. For example, Alabama protects the title "engineer" as well as the title "professional engineer."

ance with local conditions. To obtain some measure of uniformity, however, engineering practitioners have developed a "Model Law," which is used as a guidepost. This law has been repeatedly revised and brought up to date since it was first drafted in 1941. The most recent revision was made in 1984. See NCEE Model Law (1984 Rev.)[20] in Chapter 4.

By the very nature of the situation, there will always exist differences in law interpretation and administration procedures. This will bar exact similarity in wording and likeness in operation of the various states' machinery for Engineers' Registration; however, some degree of uniformity is a goal toward which all states are striving.

SERVICE TO THE PUBLIC

The work of the engineering profession, more than any other, truly concerns the public's safety, health, and property. We must look upon this from the long-range point of view, in matters of service to the public, acceptance by the public, and recognition by the public. Only thus can an increase in the public's respect for the engineering profession take place.

Responsibility to the public, which is implicit in our state registration laws, is focusing the attention of the practicing engineer on the engineering profession; a further effect is to bind closer together all technical branches of engineering through this common requirement. Thus, to the public, the engineering profession can present one common front, one united profession. Practically every design, every operation, and every process undertaken by engineers has public implications. Almost the entire range of engineering activity concerns the public health and safety, and competence to perform with requisite knowledge and skill is essential to the public good. If the engineer builds a bridge, designs a dynamo, constructs a steam turbine, or develops a chemical process, he or she must keep the safety and well-being of the ultimate consumer, the public, uppermost in his or her mind. If the project is to build a water-supply system, install an air-conditioning system, or light a building, the public's health and safety are involved. The design and construction of buildings, the planning and operation of waste-disposal plants, the building of an engine or an airplane, all involve the safety of the public, and many of them involve the public health.

PROFESSIONAL RECOGNITION

Engineers are pioneers, pathfinders, trail blazers in almost everything they do—with only one exception. They have left it to the other learned professions to be the pioneers in the field of securing professional recognition. Over 200 years ago, the medical profession began to seek professional recognition; the legal profession, about 100 years ago; but the engineering profession has been laggard, and only in the 1940s and 1950s did it begin to look up from its preoccupations and direct its planned activity to the problems of professional standing and recognition.

In an article in the Michigan *Professional Engineer*, in 1951, Dr. Melvin Nord, P.E., wrote:

> In the last analysis, however, the principal factor in the lateness of the development of engineering registration has been not so much that people are unaware of engineers, as that engineers are unaware of people, including themselves. That is to say, the typical engineer thinks of engineering, but does not think of engineers. Since no one else does either, we can see why engineering registration has lagged, until recently at least.

In 1986, licensed individuals in engineering work numbered about 343,000, while the total number of practicing individuals was believed to run in the order of 1,627,000. See Table 2.2.

One of the principal handicaps in the public's understanding of the engineering profession has arisen from the promiscuous use of the term "engineer" to designate a variety of occupations such as steamboiler operators and locomotive drivers, as well as skilled designers and planners of buildings and structures and mechanisms, brilliant research workers, eminent consultants, and operating executives. Incidentally, those who still have doubts about the necessity or desirability of engineering registration laws should just speak with members of their state board of examiners about their experiences in reviewing applications for registration. Even the most skeptical would soon be convinced of their necessity.

Protection of Title

Not only are engineers' registration laws necessary for the safety of the public and for the protection of the public interest; they are necessary for the protection of the good name of the profession. This is because a profession is judged by the failures of the incompetents* and the conduct of the unworthy, unless a clear dividing line is established in public

---

* Ross, Steven: *Construction Disasters—Design Failures, Causes and Prevention*, McGraw-Hill (1984).

TABLE 2.2

	Medicine (1982)	Dentistry (1983)	Law (1980)	Engineering (1984)
First licensing law	1760	1841	1890	1907
First college course	1765	1840	1779	1828
Accrediting of curriculum by professional societies	1907	1909	1921	1936*
National Board of State Examiners formed	1892	1883	1931	1920
Founding of professional society of licensing practitioners	1847	1859	1878	1934
Licensed individuals	522,500	150,000	542,205	343,000†
Practicing individuals	522,500	150,000	542,205	1,627,000

* Programs were accredited by the forerunners of NCEE [the National Council of State Boards of Engineering Examiners (NCSBEE)] prior to 1936. Also ECPD (presently ABET) was involved prior to 1936. Accreditation actually came about because of discrepancies in individuals' engineering education and concern by state boards that a college degree in itself may not entitle a person to practice engineering because of the lack of uniformity in educational standards.

† The figure for licensed engineers is approximate since so many engineers were registered in so many different states that the actual total is difficult to ascertain.

SOURCE: FOR ALL PROFESSIONS EXCEPT ENGINEERING, *Statistical Abstract of the United States, 1986,* U.S. Department of Commerce, 106th ed. For engineering: National Council of Engineering Examiners (NCEE).

recognition between the lawful practitioners of the engineering profession and those who are not permitted to practice.

The pioneers of Engineers' Registration fought for the protection of the title "professional engineer" because no other agency could be expected to accomplish this objective. They foresaw that without registration laws, there was no way to stop the practice of engineering by the nonengineer. By means of Engineers' Registration, the profession is in a position to enforce its code of ethics and fair practice. Otherwise, such documents would be mere scraps of paper, of no real value to the profession.

We saw why engineering must be regulated by the state. One way such regulation is achieved is by protecting the use of the title "professional engineer." This method has been declared constitutional by our state courts and is being followed rigorously by our registration boards and national and state societies of professional engineers. The abuse of the designation "professional engineer" is being checked, and the incompetent and the unworthy are being halted in their attempts to become registered as professional engineers by the established requirements of experience and examination by our state examining boards.

Without registration laws, engineers would have been restricted in their now established right to practice their chosen profession. In the early days of the registration movement, legislation sponsored by architects would have eliminated or subordinated the engineer in the structural field. It is a matter of record that "physicians endeavored to monopolize the sanitary field, and that accountants sought to exclude others from making financial statements and reports." The pioneers of Engineers' Registration fought off legislation proposed by lawyers to deprive engineers of the right to prepare contracts and similar documents and to engage in arbitration proceedings. Even real-estate brokers endeavored to curtail the right of engineers to make appraisals.

Naturally, the esteem and prestige many outstanding engineers bring to the engineering profession, as well as that which is inherent in the profession, attract many—some of whom are qualified and some of whom are not—who want to use the title "professional engineer." These people may be good citizens and neighbors, but many are not in possession of adequate technical knowledge.

For instance, there are on record amusing examples of interesting "new"—but potentially dangerous—concepts of physical phenomena put forth by such untrained people. From one of them we learn that the flow of alternating current in a circuit is retarded by "impotence." Most of us know that something does resist the flow of alternating current, but we call it "impedance." From another would-be engineer comes the thought that "moment of inertia" of a structural section is a matter of momentum or flywheel effect. Others state that "modulus of elasticity" and "section modulus" are one and the same thing; that "bending moment" is that instant when a beam begins to bend.

It is because of these farfetched notions, the ignorance from which they spring, and the dangerous potentialities linked thereto, that it is necessary to have minimum legal standards for those who would present themselves to the public as professional engineers and to regulate the actual practice of the profession.

A few more examples of applicants for licensure who have been rejected will further impress upon us the need for Engineers' Registration, with its attendant power to control. The following examples were provided by an outstanding and well-known former member of a state board of examiners.

An applicant appearing before the board reflected strong evidence of a limited education. He presented an application with an experience record listing many years of impressive professional engagement on "the design of mechanical equipment of vessels of the United States Navy." At the oral interview, he was asked the most elementary questions:

QUESTION: What is a horsepower?
ANSWER: 33,000 pounds.
QUESTION: Do you mean that a horsepower is the same thing as 16½ tons?
ANSWER: No, sir. It is twice as much.
QUESTION: Wait a minute—16½ tons equals 33,000 pounds, and you stated that a horsepower is equal to 33,000 pounds. Then what is the relation between a horsepower and 16½ tons?
ANSWER: They are the same thing.

Another applicant claimed extensive experience in the design of heating and ventilating and power equipment. When an elementary question on heat calculations seemed to cause confusion, the examiner tried to simplify matters:

QUESTION: One pound of water, when evaporated, will yield how many pounds of steam?
ANSWER: I am sorry, sir, but I really have no idea.

The candidate questioned in the following example claimed to be a heating and ventilating engineer:

QUESTION: How much heat does it take to make a pound of steam?
ANSWER: It takes 33,000 Btu per pound of pressure per minu' [illegible] eat transfer.

Another applicant without engineering education styled himself an "amusement engineer" and claimed to be the leading expert on the design of roller coasters. He had "designed" the huge roller coasters in some of our principal amusement resorts and for large fairs and expositions. He was questioned about his use of formulas:

QUESTION: How do you calculate the stresses in the structure?
ANSWER: I know from experience how large to make all the pieces.
QUESTION: What calculations do you make?
ANSWER: I have a formula in my office for the force of a car going over a curve.
QUESTION: Do you employ any other formulas?
ANSWER: No, sir. That is the only one I need.
QUESTION: What factor of safety do you use?
ANSWER: About 90 percent.
QUESTION: Do you mean that your factor of safety is less than unity?
ANSWER: Yes, sir, less than unity.
QUESTION: Why do you do that?
ANSWER: To give the customers a bigger thrill. They have to hold on when they go over a curve.

QUESTION: Isn't that rather dangerous?

ANSWER: No. When a car if going fast, you don't need 100 percent safety. You get across before anything happens.

Protection of the Qualified

The individual engineer has invested many years of her or his life in professional education and has gained hard-earned experience in the field in order to qualify as a "professional engineer." The engineering profession itself has invested decades of united effort to win public esteem and recognition for those who bear the legal designation of "professional engineer." All this effort would be to no avail if the use of the term "professional engineer" were left unprotected.

The term "professional engineer" has too often been misppropriated, with resulting abuse to the dignity and respect that is the right of the qualified engineer. The registration law protects qualified engineers in their rights to practice their profession, against restriction, encroachment, and unqualified competition. The unauthorized use of the term "professional engineer" in publicity has been almost eliminated. In one year in New York City, 105 listings of unregistered persons as professional engineers were eliminated from the telephone book, and 217 similar violations in advertising and the like were corrected without the necessity of court action.

The public must be able to differentiate the incompetent from the competent and deserving. They must be able to distinguish quickly and easily between the lawful and the unlawful. Enforcement of the registration laws provides this protection.

The results of not protecting the title "professional engineer" have already been experienced by boards of examiners for registration. This difficulty has shown itself largely with state agencies that operate under the ruling that only registered engineers may get salary increases or promotions. This ruling at first glance appears sound; but for many of the individuals involved it becomes a serious matter, because, though classified as engineers by the state agency, they are not engineers within the legal meaning of the term and are not doing engineering work. It would be better if capable people were given titles describing the work they are actually doing, rather than being classified simply as engineers. It would seem to be a simple matter to have these titles changed to suit job operations, but the persons operating the agencies in question are unwilling to do so. They want the boards to find some dodge or excuse for registering the individuals involved as professional engineers. As a result, what do we have? We find a few state agencies shouting loudly for high-quality work and advocating civil-service examinations as a means of getting that quality and at the same time working to weaken and in some cases to tear down present board standards for registration.

Professional Consciousness

Engineers' Registration has been a strong force in developing professional consciousness in the engineer. Because registration is a legally recognized testimony of competence, it protects the engineer and it protects any who use the engineer's services. Setting minimum standards for the professional weeds out incompetents and undesirables and shields the engineer from unfair competition. This pays off economically, as well as in recognition and protection. Thus, registration automatically nurtures a professional outlook, which invariably brings a long-term gain.

Professional consciousness is one of the essential qualities of a professional. It is a feeling of loyalty, a sense of belonging to a group, a quality that sometimes finds different expressions in the junior, the young graduate, and the senior engineer. This difference, however, is not a fundamental one, but seems to be the result of differences in age, experience, and wisdom.

When professionally mature engineers comtemplate the advantages of Engineers' Registration, they tend to consider with a true sense of value such aids to professional development as:

1. Eligibility to perform professionally and to do creative work in engineering.
2. Pronounced ability to apply engineering knowledge.
3. Legal recognition in the chosen field of work by a board composed of the professional's peers.
4. Improved professional status: engineering registration is evidence of ability and certification of individual competence according to recognized standards.
5. Job advancement possibilities: many engineers find that their companies recognize and reward the achievement of professional engineering status.
6. Defined legal status: in court cases, the fact that registered engineers are in charge of the project can mean the difference between winning or losing the case. A contract to do engineering work is not legal unless the engineer is registered.
7. Improvement of the profession: Engineers' Registration tends to unify

engineers on a professional level, sets a minimum standard, and increases public recognition of the professional standing of engineering.

On the other hand, when young engineers, upon graduation, approach the question of attaining Engineer-in-Training (EIT) status, they are likely to ponder their recent experiences and, against the background of the struggle and sacrifice of their engineering education, to reflect on the cost of certification as an EIT and the nature and validity of professed returns and ask, "What would an EIT certificate do for me?"

It seems to be part of the eternal nature of youth to display a remarkable appreciation of immediate advantage, to manifest impatience with slow, even though steady, personal progress, "to take the cash and let the credit go," and generally to underestimate those great intangible values which to the mature and older mind stand out obviously as the best foundation of a sound professional basis.

The individuals who elect to become engineers-in-training will be the recipients of the major objectives of a program designed:

1. To make them more conscious of their obligation as members of the engineering profession, and to acquaint them with the procedure, requirements, and advantages of registration.
2. To permit them to take the first step toward registration by passing a written examination on theory while the subject matter is fresh in their minds.
3. To assist them in obtaining immediate professional affiliation, guidance, and ultimate protection.
4. Through increased professional consciousness, better knowledge of the registration procedures, and reduced fear of the final examination, to encourage young engineers to acquire professional status by registration as soon as they are legally qualified.

Technical Development Improved

Another effect of Engineers' Registration is that the technical development of the individual engineer is being improved. Through the influence of Engineers' Registration and to meet its needs, the program of accrediting engineering programs in colleges and engineering schools was inaugurated through ABET. This in turn is resulting in a rise in standards of engineering education in matters of improving curricula, quality of teaching, and educational plant and equipment.

Before Engineers' Registration, most engineers were "self-made."

They learned their jobs through experience, and many of them turned out to be good engineers; however, with changing times and through the effective instrumentality of Engineers' Registration, professional education, graduate study, and advanced degrees have now become pretty much the rule as preliminaries to a professional career. Now, although there are forces afoot to restrict the obtaining of a license to graduates of accredited programs in engineering, in a number of states, the gates to registration are still open to the person who has not had the advantages of a formal engineering education. The applicant is provided with the same opportunity as an engineering college graduate to become registered by sitting for the same written examination; however, the applicant must have met the same additional criteria which are required of the engineering graduate.

INDUSTRIAL TREND

There is a trend in industry today which sanctions the policy of reserving the title "engineer" solely for one qualified by registration. All industries are becoming aware of, and alert to, the desirability of having registered engineers in their employ. Such a policy has definite public appeal and affords some protection to the organization. A registered professional engineer in charge of design is good insurance against possible charges of criminal negligence, should equipment—from electric toasters to bridges—cause injury or loss of life or property.

To licensed engineers, registration is of obvious importance. To personnel or employment departments of large engineering organizations, the effects of registration can be measured without prejudice. It is well known that such groups consider registration as advantageous to the public; however, there are others who feel that their own methods of selection, weeding out, and training will assure them of the competency of their engineering staffs. Moreover, there is a not uncommon feeling persisting that registration can add little to graduation from an accredited engineering curriculum in assuring competency. In such cases, of course, the aims and ideals of Engineers' Registration are not fully understood. In addition, it is easy to see that nonengineers in many private industries are unable to observe the influence of registration upon the competency, efficiency, or standards of the engineers in their organization; however, most feel that registration is gaining public favor.

PROFESSIONAL UNITY IS NECESSARY

Through our registration laws, the principle that engineering is one profession has been established, although specialities may be many. Law and medicine have as many specialities as engineering, but lawyers and doctors would never consent to the legal subdivision of their professions. Whether individuals write E.E., M.E., Chem. E., or C.E. after their names, they possess fundamentally the same basic educational training, the same method of analytical approach to technical problems, the same governing professional qualifications, the same ideals of professional practice, and the same interest in the profession's standing and reputation.

Because registration is only comparatively slowly being required of those practicing engineering as employees in industry and the workplace or those engaged in engineering works, the engineering registration figures do not accurately reflect the size or complexion of the engineering profession as a whole. Consulting engineers and engineers employed on public works generally must be licensed, and the engineering registration rolls, therefore, are strongly flavored with those particular branches of the profession.

Engineers must strive toward professional unity and recognition if they are to upgrade their profession. That should go without saying, but many persons engaged in engineering works appear to be overlooking the ready-made means of strongly establishing the foundations of their profession and of bringing all engineers together.

If it is true that there are presently over 1 million engineers in this country, then it takes little imagination to ascertain what magnitude of powerful combined influence could be exerted, which at present is too dissipated to be effective. Naturally it could be made effective if every person who uses the title engineer were officially and legally recorded as a registered professional engineer.

3

WHAT IS PROFESSIONAL ENGINEERING?

What is professional engineering? Who has the legal right to call themselves professional engineers? There is only one answer, supported by the police powers of the state, and that requires registration with an official board of examiners, in some states called a board of registration.

We have seen heretofore that the only legal basis for registration is for the protection of the public. But how is the public to know what is professional engineering and who is a professional engineer in good standing with a board of examiners?

All our states and territorial possessions have statutes regulating the practice of professional engineering. A typical statute states:

> The term professional engineer . . . shall mean a person who engages in professional engineering (before the public) as practicing in the rendering of a service or creative work requiring education, training, and experience in engineering sciences and the application of special knowledge of the mathematical, physical and engineering sciences in such professional or creative work as consultation, investigation, evaluation, planning, design, or responsible supervision of construction in connection with any public or private utilities, structures, buildings, machines, equipment, processes, works, or projects, wherein the public welfare, or the safeguarding of life, health, or property is concerned or involved, when such professional service requires the application of engineering principles and data for the purpose of securing compliance with specifications and design of any such work.

Notwithstanding all the definitions of professional engineering and engineering alone set forth in the various statutes, drawing a sharp line of demarcation between what *is* and what *is not* professional engineering presents difficulties. If the definition covers too wide an area of engineering activity and accomplishments, it may become so vague as to cause the statute to be declared void because of uncertainties. On the other hand, if the statute is too specific in describing an engineer's functions, it becomes unfairly restrictive, with attendant confinement of engineering to its oldest branches. Indeed, this is no easy problem to solve.

PUBLIC MISCONCEPTION

The public has a definite misconception and a poor understanding of the meaning of the term "professional engineering." Some of this is due to the lack of publicity given to the profession by engineers themselves. Some of it is also the result of a sense of false modesty reflected in the thinking of too many engineers.

When talking to some people, you can readily note that they see no difference between an elevator operator and the designer of all the intricate mechanisms that go into an automatic elevator to provide good and safe operation. A number of years ago, Ohio University conducted a survey to find out what the public thought about engineering.* The occupations of the respondents represented nearly every economic and social level in the United States. Some of the occupations listed were: teacher, lawyer, doctor, engineer, factory worker, school superintendent, waitress, journalist, housewife, secretary, clerk, laborer, merchant, and banker.

In a personalized random sampling, the question "What is an engineer?" was asked. A number of surprising answers were received.

> An engineer is a normal person who is more than normally interested in what seems to be a dull job to an outsider.
>
> His tools are his brains and his hands, but he works without a well-rounded personality.

Other startling answers were

> A well-rounded person, but sometimes one-sided.
>
> Engineers are too wrapped up in their work to have normal social life.

Engineers were described as more serious individuals than others, tending not to be able to look up from their preoccupations and comprehend the world about them.

---

* "Ohio University Man-on-the-Street Survey. . . . " *American Engineer*, December 1956.

Have you ever heard a carpenter disclaim any skill in wielding a hammer or saw? Have you ever heard a butcher depreciate his or her value to anyone in the ability to get the best "cut"? No, where skill is of paramount importance and is the essence of a person's calling, professional pride is much too fierce for such self-disparagement. In spite of that, we often hear persons who profess to be engineers disclaiming any specialized knowledge. Some of them not only admit their lack of skill in working with the essential tools of their profession (mathematics, science, etc.), but even appear to glory in this lack and cheerfully confess ignorance of the basic mathematical techniques. They will often depreciate themselves in the eyes of those around them by boasting that published papers and technical articles in the magazines are too "highbrow" for them to understand.

Some of us who have the opportunity to associate with engineers from schools all over the country have found that a startlingly small percentage of them consider engineering to be a profession. They come to work in old clothes and generally act like laborers. During their off hours, they do not study or try to better themselves culturally or technically. Perhaps our educators have neglected to tell their students what being a professional involves.

We are aware that the word "engineer" is probably misused more than any other word describing the manner in which a person makes a living. One source of difficulty in defining "engineering" is that the engineering profession is legally unable to preempt the use of the word "engineer," because "stationary engineers," "marine engineers," and others working on engines (such as "locomotive engineers") adopted the name first.

Apparently, the solution is to use a slightly different term to describe the profession, i.e., "professional engineer." Still, in some quarters of industry and education, we hear the expression "professional engineer" for one not legally registered as such.

GOVERNMENT DEFINITION

Perhaps much more must be said and written in an attempt to educate the general public about what professional engineers are and for what they stand. Unfortunately, even with the mountains of printed pages on the subject, the desired results have not been fast in developing. From some quarters, we hear the accusation that a good deal of the confusion stems from the fact that there is no definition of professional engineer that is comprehended and agreed upon by the engineers themselves. For the profession to realize full appreciation from the general public, a definition not only must be developed but, in addition, must be agreed upon, publicized, and made comprehensible to all parties concerned.

The definition of professional engineer adopted by the Salary Stabilization Board (SSB) is one showing marked similarity to the definition given above from a typical statute.* However, it is too all-encompassing and tends to weaken the results of Engineers' Registration. It places all engineers on pretty much the same footing, without regard to minimum requirements. Nevertheless, because the matter was brought before Congress, many members of Congress now have a much better understanding than ever before of what engineers do and of their importance to the national economy.

Until the present registration laws are changed to include not only those professional engineers employed by consulting firms but also those engaged in professional work not now requiring registration, the SSB is realistic. Registration by a duly constituted board of examiners will be accepted by the SSB as unquestioned evidence that an individual is a qualified registered professional engineer. There could be situations wherein the others defined in interpretation 12 (interpretation of the definition of professional employee) would not be permitted to make their services available to the general public. Only a person registered by a board of examiners is permitted to provide engineering services to the public, that is, to hang out a shingle.

The metamorphosis from a partially registered national engineering force to a completely registered engineering force must be perforce a gradual one, indeed. The day will come when all engineers will normally go through the process of becoming legally identified with their profession by licensure. As licensing procedures are strengthened and simplified and more engineers become profession-minded and learn how to go about the procedures of licensing, the practice will become widespread. This is the nature of the American engineer. Any coercion applied will only work hardships on those who have a constitutional right to claim professional classification, but for one reason or another are not registered. At present, Engineers' Registration is permissive, as it should be, and any unregistered engineer may prove that she or he has the minimum requirements by applying for registration and sitting for the written examination.

Substandard Registered Professional Engineers

Passing the required examination in no way and by no means determines the grading of engineers, but simply furnishes evidence that the

* Also see the definition in the Model Law.

registrant meets certain legally prescribed minimum requirements. Unfortunately, there are still some older individuals educated as engineers who are now high-class technicians of ability. They have lost the creative touch they once had and haven fallen into mediocre and routine jobs.

Grandfather Clause

There are still many working engineers who were admitted to the legal practice of engineering under "grandfather" clauses. Perhaps some explanation of the meaning of the grandfather clause will be helpful and shed some light on the practice.

Whenever a field of endeavor becomes subject to statutory regulation for the first time, it is essential to recognize that there are many already practicing successfully in that field. Therefore, it is necessary to accept those who apply who are satisfactory. That part of the registration law which makes this possible is known as a "grandfather clause."* Such a clause is always included in an initial professional statute. It usually expires one year after passage of the statute, subject to the discretionary power of the board of examiners. Sometimes it is effective for a longer period. For example, there was such a clause in the New York law, and it was effective until 1931, ten years after the law was passed in 1921.

We will always hear some unregistered persons who think they should be registered tell us how much better they are than some of their registered associates. Our state boards will always give such people a chance to prove their claims by sitting for the written examination.

WHAT PROFESSIONAL ENGINEERING IS NOT

Engineers' Registration is not an elite country club for the eminent practitioner. The boards do not go out of their way to set roadblocks for the license-minded engineer to hurdle. The boards want individuals to get licenses, if they have the professional experience and professional know-how. Our younger engineers have the same legal advantages as their older and more famous colleagues.

Engineers' Registration should not be thought of primarily as protection for the licensed engineer, but rather as protection for the public, although it does protect the engineer as well.

Most supporters of professional Engineers' Registration say it raises

---

* The grandfather clause has expired in most if not all the states at this time.

the standard of practice; and it does, because those who are not competent cannot call themselves registered professional engineers. The eminent are not affected; it does not raise the level of normal competence. But it does act as an incentive to the younger engineer to achieve recognized legal competence at the earliest possible date.

Many of the objections to Engineers' Registration have been answered by the passage of more and more laws. Many other objections will, with the passage of time, pass into oblivion, just as soon as the basic function of professional enginering is understood to be the responsibility of the individual, whether self-employed or in the employ of private industry. This is a prelude to an impending situation wherein all engineering practice will be carried on only by those who have legal status as registered professional engineers.

WHAT PROFESSIONAL ENGINEERING INVOLVES

The work of the engineer is primarily creative, aimed at making the world an ever-better place to live in. In the beginning phases of construction, in the planning and design phases of a project, professional engineers make their presence felt.

In Design and Construction

When design is involved, the various state registration laws require that before the design can be filed with the local authorities the designing engineer must affix her or his seal thereto, for record and to fix responsibility. This in in keeping with the police powers of the state, to guard against incompetents and provide for the safety and well-being of the general public. Design of construction projects and final construction require the employment of engineers especially qualified by training and experience in the various branches of engineering involved, such as civil, mechanical, electrical, structural, industrial, chemical, sanitary, and heating and ventilating.

An engineer, when in responsible charge of the field construction, such as resident manager or field superintendent, must see to it that the work is performed strictly in accordance with the plans and specifications. The engineer may be called upon to make on-the-site decisions, but in some cases will enlist staff help from the office wherein the design originated. In some states, the construction engineers in responsible charge are now required to be registered in the state in which the project is being built.

And since their work takes them from state to state, multistate registration becomes a problem, one which we consider in a future chapter.

In Industry

The professional engineer in industry plays an important role in the scheme of things and makes a definite contribution to society. People are almost completely dependent upon the products of engineering know-how, from the moment they arise in the morning and use the toaster to the moment they retire and pull the light cord or throw the switch.

The professional engineer in industry endeavors to hold a place in the ranks of management. Since the work is of a creative nature, the engineer's knowledge and experience are used to improve upon existing processes and to promote progress. Management looks to the engineer to reduce production costs with attendant safety. Plant operations are looked upon from the safety angle, and preventative maintenance is replacing the familiar mechanic with oil can and grease gun. No matter how much standardization is achieved with respect to machines and methods, the engineer's work cannot be standardized, for it is the very lifeblood of industry.

Because of the necessary close association with management on one side and labor on the other, the engineer is well aware of the aims and problems of labor, as well as the wishes, desires, and problems of mangement.

The scope of an engineer working in industry has a tendency to become limited, as compared with that of a doctor or lawyer. Professional engineers can enjoy greater horizons of thought and association if they look up from their daily tasks and avail themselves of the opportunities to combat narrowing influences. Should they allow the situation to go unchecked, the once-energetic, mentally active young engineers can become so preoccupied with immediate problems that they become a total loss to their profession and to society.

As members of a technical society, engineers become acquainted only with other engineers interested in their particular fields. As members of a professional society, such as the National Society of Professional Engineers, they may participate in the overall activity of their local chapters and feel the pulse of all fields of engineering endeavor—and on a professional level. There is much they can do there. Among many other things, they can raise their voices to help bring about better public understanding of professional engineering, and thus improve the economic status and social welfare of all engineers.

In Government

The functions of government include, among others, developing and conserving natural resources, providing means for the economic growth and progress of the community, and looking after the general well-being and safety of the citizenry.

The Nuclear Age is making, perhaps, its greatest impact on engineering talents and resources. The professional engineer has a key role in the tremendous technological development that began with the opening of this new scientific frontier just a few years ago. The same may be said of the latest satellite and computer developments.

The professional engineer of tomorrow not only will fall heir to impressive accomplishments of the past but will be required to develop a greater proficiency than his or her predecessor and a greater awareness of the economic and social impact of his or her works upon others.

Government service offers the future professional engineer singular opportunity for professional advancement. There is prestige and responsibility in this calling for the engineer, for it has far-reaching effects on the health, welfare, standard of living, and economy of the nation.

One word of caution: the professional engineer runs a greater risk of becoming limited in scope in government service than in most other callings. The remedy, however, is the same as we proposed above for the engineer in industry.

In Consulting Practice

Of the many fields of endeavor that interest the professional engineer, perhaps consulting practice is the most satisfying and inspiring. In the fullest sense of the word, the engineer in a consulting practice enjoys the thrills, hopes, and fears of an independent businessperson. Satisfied clients are the best advertisement, creating goodwill. This is the one field in which an engineer must be registered in order to offer services to the general public.

Every assignment is just a little different from the last. This makes for interesting work and provides for a wealth of experience rarely equaled elsewhere. There is another satisfying attribute, the respect generally accorded the consultant. The consultant's advice is sought and appreciated at much the same level as the doctor's and attorney's. The consultant develops a natural interest in public life by coming in contact with businesses of all sorts and professions of all kinds.

Because of their familiarity with technical aspects of the projects on which they are engaged, because they have worked in the field and know how the work will be done, because they have planned and directed work, and because they understand the economic problems as well as the technical problems, consulting engineers are most suited and able to appraise the problems and scope, possibilities and limitations of new enterprises.

Consulting engineers make available the skills of the professional engineer to the public. Their professional engineer's licenses permit them to practice legally. They have proved themselves of minimum competence through the medium of state licensure. They have, in effect, sworn to look to the safety and welfare of the public.

In Research

The ever-increasing importance of research to the economy and security of the nation is being demonstrated in many ways. With the almost complete dependence of our entire economy on science and with the outstanding accomplishments of research people, the role of the research engineer becomes more and more apparent in the thousands of new products, devices, and processes appearing in our lives. It is expected that research groups, within the next decade or two, will produce technology as an industrial product of their own, and the competition for engineers and ideas will reach major proportions. Like other engineers of tomorrow, the research engineer will differ in many significant ways.

Professional engineers in research will differ in the organization of their activities, their educational preparation, and the problems they will have to solve. Much will be done to control the environment and to predict and control disasters, such as floods and hurricanes. Transportation methods will be changed to improve safety; pilotless aircraft will be commonplace. The solution to multiple business problems will be tackled with engineers' new knowledge of the applications of computers. Better-planned production and improved forecasting and control of inventories will be given greater impetus by their knowledge of the principles of servomechanisms and feedback circuits.

Tomorrow's research engineer will require much more extensive education and more fundamental training than are required today. The young undergraduate engineer will have to make a much earlier decision between the "research option" and the "engineering design option." We may witness a gradual departure from the classical concepts of engineering research fields as we know them—electrical, mechanical, civil, chemical. The training of the research engineer of tomorrow may resemble that of a physicist, with much more basic knowledge of mathematics and nuclear- and solar-energy applications.

From the financial standpoint, the professional in research has unlimited possibilities, if she or he is well equipped with the right aptitudes and personal characteristics. Research can be as remunerative and gratifying a career as any that can be foreseen at the present moment.

In the Armed Forces

It's well known that for centuries the basic art of warfare remained static. Strategy and tactics, together with the sheer weight of the size of the army were, up to a relatively short time ago, the deciding factors in winning wars. Now this art has been transformed into a science by the application of science and technology to weapons and their uses.

While the research engineer thinks up new ideas and researches them, the professional engineer has the task of transforming those ideas into real products or practices. The engineer's know-how is the important ingredient that catalyzes advancement.

In the fields of engineering development and production, the professional's knowledge and experience must be put to good use. As programs within the armed forces continue to grow, more and more engineers in industry will find their talents utilized without having to leave their current employment. Armed forces development work, although done to a great extent by industry, will require a hard core of researchers, scientists, and technicians within the military establishment to introduce and define the areas in which the work will be performed. Also, this group will be required to coordinate the thinking within and without the military organization.

The nation's technology needs to be encouraged and effectively utilized, for it will continue to be a big factor in our national defense system. Professional engineers in the armed forces program of scientific and technical endeavor will need to expend infinite effort on the research and development front, the production front, and the combat front. The professional engineer is essential to the avoidance of conflict, for survival in conflict, and for contributions to a better civilization.

In Engineering Teaching

In many states, engineering teaching is considered professional engineering and a license to practice engineering is required to teach advanced engineering courses. Most states give some credit for engineering teaching as engineering experience for any academician who decides to pursue licensure as a professional engineer. See Chapter 8, "Teaching as Qualifying Experience."

THE ENGINEER AS AN EXPERT WITNESS

When we go into the courtroom, we leave our laboratory and go into the laboratory of the attorney. There we will be discussing technical information that is beyond the general knowledge of the lay public. In fact, it is that lack of general scientific knowledge that leads to the real purpose of expertise. As engineers and scientists, we know, with scientific accuracy, certain things that the general layperson can only believe or suspect. The introduction of such knowledge as engineering fact is not an invasion of the province of the jury. Actually, it is introduced as an aid to the jury in explaining the difference between scientific fact and lay opinion. This is the first and primary function of the expert witness. The second function is that of explaining the basic facts of science and engineering as applicable to any particular problem.

The extension of these data into the area of opinion is proper when the foundation for such expression is properly based. It is here that we run into a portion of our dilemma; frequently we find that attorneys do not develop their own understanding of the scientific data to the point where they are able to make the fullest use of expert witnesses. This makes the expert's job doubly difficult in that it prevents her or him from answering factual questions for the benefit of the court and the jury simply because the engineer's attorney does not know what to ask.

To demonstrate the difference in thinking between the legal profession and the engineering profession, consider a black and white color chart. The legal mind, stretching through the entire spectrum of black and white, would have a very small portion of black at one end and white at the other. The center portion of the spectrum would be almost all gray because the thinking and the technical approach of the legal mind stays in the areas of the grays, representing the wide variety of ways to approach legal issues. Legal thinking must constantly change as new decisions, new laws, and other similar changes come into being. On the contrary, engineering thinking, since it is primarily based upon scientific fact, and would have very little gray; most of the spectrum would be black or white.

The dilemma that we face is the reconciliation of these two spectrums, making it possible for the engineer to communicate with the legal mind that the facts being expressed in the engineering concept are factual and not just opinion. To add to the dilemma, many terms used in engineering have become popularized and to some extent distorted in everyday usage. As an example, the word "pound" is used by the layperson to mean weight, whereas in engineering it can also mean a force. Obviously the intermeshing of such terminology creates discomfort, at best, in communication between members of the two professions; at worst, it produces disbelief on the part of the legal mind for the scientific explanation that uses the term in its proper scientific context. Other examples include the common usage of momentum, mass, and similar words that mean one thing to the attorney and an entirely different thing to the engineer.

In addition to the misunderstandings generated by scientific versus popular definition of words, there is the difficulty caused by the use of equipment or data based on averages rather than specifics. An example is the Breathalyzer (a breath analyzer). It is well designed scientifically and, used properly under adequate controls and with proper maintenance, it is capable of analyzing the vapor submitted to it for analysis. However, there is a very interesting assumption in the use of this machine that frequently causes improper results; the designers have assumed the relationship between blood alcohol and breath alcohol to be constant. This assumption has been made even though many researchers have found that this relationship is far from constant. When we consider that a prison sentence may result from such an assumption, it is obvious that the dilemma of the scientist and engineer working with such data is serious. Challenging the results is difficult because of the support the Breathalyzer enjoys from some portions of the scientific community. The fact that this disagreement exists should cause a rejection of its use in cases involving the freedom, security, or earning capability of an individual charged with any kind of crime. Thus, the dilemma grows.

We deal in reconstruction with the basic physics, chemistry, and mathematics that are classical in development. They are not assumptions; they are supportable scientific facts. Nonetheless, these facts are frequently rejected by the legal discipline because it does not understand them or because it suspects that the scientific community is simply not capable of determining the facts. These legal minds arbitrarily throw out the data and results derived by scientists, yet accept guesses from untrained or prejudiced individuals in scientific matters.

The situation can be demonstrated as follows:

When we make calculations pertaining to the stopping distance of a vehicle, we are using physics and mathematics in relating the amount of work done by or on an object, in relationship to the kinetic energy of that object. There should be no question about the accuracy of such calculations, provided they are properly founded upon equitable and proper engineering. Such a foundation is not difficult to present, but, unfortunately, the legal parameters traced around such studies are such that variables that have absolutely nothing to do with the real problem are introduced. Since these variables cannot be measured, the entire calculation is rejected. If the legal parameters are fully explored and developed and the testing fully explored and developed, there should be no reason to reject this evidence. Calculations relating to chemical or physical actions, results of testing, and results of examination conducted by qualified individuals should not be rejected if they are based upon proper scientific studies, proper scientific appoach, and proper presentation of qualifications.

It appears to be surprising to some students of the law, attorneys, the judicial group, and the legislative body that reconstruction of accidents (autos, fires, explosions, etc.) can be conducted with remarkable accuracy in many instances. The expense of reconstruction will vary from problem to problem, but in each case there is almost always some portion of the problem that can be answered by scientific presentation.

Engineers find it difficult to understand the concept of rejecting evidence about the direction of forces involved in an auto or other impact-type accident beause they know that vector forces can be accurately measured by study of the units involved.

It is interesting to note that the "moon landings" and "moon returns," requiring thousands of miles of communications and extremely accurate radio control, were based on the same physics as is used in our reconstruction work. The entire moon program clearly established and proved the validity of classical physics. It would seem that such a proof would allow the admissibility of reconstruction evidence in court. It should be noted, in fact, that when reentry was planned for one of those trips, the calculations were made and the reentry established while the spaceship was on the far side of the moon. In fact, the actual shooting or ignition occurred in that location aiming for a target some six to five miles square at the upper reaches of the earth's atmosphere and usually ended up in a miss perhaps as small as ½ to 1 mile. Our difficulty becomes one of not understanding how the laws of science that allowed this degree of control can be rejected by courts when autos, buildings, or similar structures fail.

Obviously, the legal profession is interested in scientific knowledge; the many courses, symposiums, and special study courses offered and accepted by the legal profession confirm this. We do not believe, however, that the problems associated with the introduction of serious physical studies to nonprofessionals have been solved. We feel that the final answer in resolving our dilemma lies not so much in eliminating the technical parts of the subject as it does in separating physical science from the language of that science (mathematics). It appears to be the "numbers" part of physics that provides the problems of misunderstanding on the part of the legal mind. Since much of the material in the basic physics courses does not need to be treated numerically, it is not only possible, but completely practical, for the attorney to study many aspects of physics without a mathematical base.

Certainly the more mathematics that an individual understands, the more easily the scientist can approach him or her; an individual who lacks such a background, however, can appreciate the fundamental science that plays such an important part in the world in which we live. As scientists and engineers, we have a duty to aid in the understanding of the physical world by using examples in which the understanding of advanced mathematics is not necessary.

While we may be able to solve a portion of our dilemma by eliminating large amounts of the mathematical references, the technical vocabulary cannot be avoided. Fortunately, physics, unlike many of the scientific areas of study, lends itself nicely to a lecture type of presentation in that it offers an almost unlimited amount of demonstration material. This material is readily available and there are numerous examples of each of the physical laws that can be used to demonstrate the physical law as well as its importance.

The classical physics of Newton is still applicable to the world in which we live. Modern physics is an extension, an explanation, of some problems that Newton did not face or, in fact, even consider. Nonetheless, Newton's physics, the classical physics, fits as a special part of the physics of today. Inasmuch as physics as we know it has been in existence since approximately 1700, it is about time the legal profession recognized that these laws are available, do work, and do in fact control all our physical surroundings.

Thus our task is one of establishing in the minds of the legal profession, in all of its aspects, that we are not guessing, that we are able to make the calculations that we say we can make, that the calculations are exact, and that the only problem we face is the improper use of these facts or lack of understanding on the part of the person using them.

It is frustrating to have an uneducated, relatively untrained individual look at a set of skid marks, perhaps even measure them in a haphazard manner, and then testify about the speeds indicated by those skid marks, and at the same time to have a calculation based on scientific analysis of the same marks rejected. It is equally frustrating to be told by members of the legal profession that they do not believe the data developed by the scientific community because "they" decide what will be influential in making the determination. These things can result in a gross miscarriage of justice.

But of all the problems faced by the expert, the most difficult one is the presentation of so-called "factual information" by unqualified people. We repeatedly are faced with comments about debris, marks, gouges, and all sorts of so-called "factual information" which, if analyzed properly, could be a valuable aid in reconstruction, but which, if based on lay information and presented by untrained people, is not. In a recent case a witness, having qualified himself as an expert, testified that he did not understand nor had he ever understood the meaning of the coefficient of restitution. Nonetheless, he made a complete reconstruction based upon velocities and momentum without ever giving consideration to the coefficient of restitution. We earnestly urge all attorneys dealing with the automobile or industrial accident to at least become familiar with the terminology and the importance of the various concepts that are applicable in such work. The value of this knowledge is twofold: it allows the attorney to test the knowledge of the hired expert, and it allows the attorney to test the knowledge of any opposing expert.

As reconstruction experts, engineers and scientists can provide and use scientific and engineering facts to arrange the physical evidence so that each part of a construction can be tested, or so that the experts can look at an already completed proof and test its essential elements. In all cases, the proof is whether or not the elements used are linked or joined by scientific fact or by ill-conceived empirical wax.

Remembering that scientific laws explain only "how" and not "why," we see that the "how" becomes a "joint" that holds together two or more parts of the physical evidence and must provide the maximum strength. When the "joints" are provided by expert testimony, test the joint by testing the expert's qualifications at the time the joining was made. Ask the expert to reduce his or her opinion to its most fundamental root or derivative in science. The true expert expects this; the would-be expert relies on tables, charts, and the empirical work of others without going to the root in science. The proper expert can produce his or her own tables and charts, but such production is almost never necessary in a specific case.

When facts are in dispute with "what happened" and provide the best available "how it happened," then we are certain of the best "why" decisions from our American jury system. Only when the jury has a complete picture of "how" and "what" happened is it able to express answers as to "why."

THE PE LICENSE—WHAT IT MEANS TO INDUSTRY'S ENGINEERS

The aura that surrounds the title "professional engineer" is intriguing. It identifies a respected community of people who rank with doctors, lawyers, and bankers in the eyes of the people of this country.

A strong sense of comradeship and solidarity exists in all professional groups, whether it finds formal expression in a bar association or in the far-ranging talk of a group of artists around a cafe table. The sense of like-mindedness which comes from associating with people of superior talents and training in the pursuit of common goals, and which puts service above gain, excellence above quantity, the rewards of self-expression above any pecuniary incentive, and a code of honor and courtesy above the competitive spirit, is one of the most dynamic of all intangible forces in modern society.

Many engineers today work for large engineering firms and feel they have no direct need for registration; however, many have obtained their licenses and there are many more who at least aspire to recognition as part of the engineering community. The effort expended to study for and pass the licensing exams and the yearly fee which must be paid to maintain the license are substantial deterrents if this license does not serve a meaningful purpose; therefore, the only conclusion which can be reached is that there are other intangible forces, which may be as important to one's associates as is a direct use in their work, which motivate them. What are these intangible forces and can an order of importance be established for them?

In their work, people are constantly motivated by the forces which are important to them, and they achieve a certain degree of satisfaction from their jobs depending upon how well the job fits their needs. The needs of professional and semiprofessional groups may be ranked according to the following in need satisfaction: creativity, challenge, mastery, and achievement:

Under *creativity and challenge* there is the need to meet new problems requiring initiative and inventiveness and the need to produce new and original works.

Under *mastery and achievement* there is the need to perform satisfactorily according to one's own standards and the need to perform well in accordance with the self-perception of one's abilities.

The professional engineering license which many engineers obtain helps to satisfy some of these needs. What are some related factors that can be established?

The possible relation and application to the job: Many engineers in industry today need a license to certify the correctness and safety of their designs. What are the implications? Even though an engineer may not need a license for the current job, the very next job may require the engineer to testify in court as an expert witness. A court of law may consider that person unqualified as an expert unless she or he is registered as a professional engineer. Engineering is nomadic in nature and no one can predict the future. The next job or project may literally take the engineer anywhere. It does not pay to gamble on one's career.

And then there is the matter of a higher salary. Most engineers indicate they chose that profession originally because of their engineering aptitudes. Very few gave financial rewards as their reason; however, the rate of pay becomes one of the most important factors in keeping the engineer contented after getting over the apprenticeship and attendant glamor of entering a new profession.

As for job security, it is one of the most important of all motivating forces, and most engineers feel they want a safe berth in industry. The PE license can help engineers obtain and retain professional positions.

The engineer strives for recognition from his or her company. If the company has any prestige value, recognition and praise by the company may motivate an engineer. A typical engineer wants to stay with a firm and wants recognition as a professional. If the engineer isn't recognized to the degree desired, the additional peer recognition made possible by licensure may provide the desired recognition.

A sense of achievement is a very basic need. In fact, this is reinforced throughout the engineer's education process each time a course grade is received. The engineer has a sense of achievement when he or she receives the PE license and becomes a member of this elite community, through meeting its rigid standards.

It thus may be seen that the forces which motivate the average engineer to obtain a professional engineer's license are the basic needs of opportunity for advancement, professional recognition, security, and a sense of personal achievement. These needs, in roughly the same order, are the prime requisites of job satisfaction for the engineer.

WHAT MAKES A "PROFESSIONAL CLIMATE"?

Professionalism is a two-way responsibility. Management must foster professional attitudes and provide recognition, and engineers must so conduct themselves as to deserve the prerogatives given them.

There are two basic ways in which a company can establish a professional climate. One is to build a program tailored to the psychological and social needs of individuals. This approach, being highly involved and personalized, requires a great amount of time and effort. Most companies, therefore, prefer the group approach.

The group approach, because it strengthens professional climate as a whole, is more practical. It satisfies needs without becoming involved in detailed individual plans, although keeping abreast of individual performance via scheduled appraisals is still necessary. Only by comparing performance against ideals can a company establish a fully professional atmosphere.

Problems of Setting Standards

Of course, there must be general agreement about what constitutes professional standards. These decisions can be difficult to deal with. Standards may be definitive statements that outline satisfactory minimum performance for various employee functions, which are defined, or may be implied in a general statement. It cannot be overemphasized that technical competence is a necessary, but not sufficient, condition for professionalism. Personal characteristics that inspire trust and confidence are also important. Standards may also be defined in terms of tangible contributions, such as patents granted, articles published, papers presented, and meetings chaired. Ideally, professional standards should involve job performance, ethical awareness, technical contributions, and civic responsibility.

There are certain relatively objective criteria for gauging professionalism, such as those by which state boards of examiners evaluate suitability of experience for licensing. They are:

1. Demonstrated integrity in professional practice.
2. Outstanding ability, which may be established by an examination of the engineer's record with respect to: (1) responsibility given by employer, (2) contributions to technical literature, (3) service on technical committees, (4) patents held and other documented achievements, and (5) achievement of advanced degrees and pursuit of continuing studies.
3. Interest in profession as evidenced by memebership in professional societies and committee work.
4. Interest in community, as apparent by involvement in service organizations.

The supervisor of engineers has a primary responsibility for nurturing professionalism. The most effective tool for accomplishing this is periodic appraisals of performance. Technical contributions (patents earned, papers presented, articles published) are more easily measured than job performance. If civic participation is encouraged, this must be done with tact, to avoid the appearance of paternalism.

Publishing should be encouraged but not required. One way of handling this is to designate a counselor who can provide advice on writing speeches and articles. This counselor should be capable of judging whether speeches and articles are technically sound and factually correct, and whether they do indeed make a contribution.

Awards Programs Promote Professionalism

Awards may or may not be financial. Cash payments for patents received, valuable suggestions, articles published, and other accomplishments can prove effective motivators. Recognition through company publications, wall plaques, and desk sets can also be effective.

One word of caution regarding financial rewards: each contribution should be judged individually for its significance. A single, set amount for all contributions would be undesirable because of the wide variation in their values. Instead, there should be a range in the amount of awards, and the specific award should be determined by an evaluation committee composed of technically competent judges.

When a company course is completed, a certificate suitable for framing could be presented. For the engineer who has recently passed the licensing examination, the appropriate ceremony could include others who have attained the same distinction, and even members of the state examining board and professional society officers.

Functional and Pleasant Working Facilities

More important than pleasant job conditions is the functional aspect of work facilities. Haphazard growth of facilities can cause inefficiency as well as inconvenience if engineers are too far away from laboratories, test apparatus, computer facilities, conference rooms, and other auxiliaries.

Should every engineer have an office? This question has been frequently raised. Nothing specific has emerged from studies of whether an engineer actually produces more in unusually pleasant, rather than in barely adequate, surroundings, assuming equally challenging assignments. Nevertheless, a private office should be provided if the nature of the work demands it. Of course, a private office can represent a form of recognition.

Other Contributors to Professionalism

Membership in technical societies should be encouraged because it provides engineers with opportunities to stay abreast of new developments in design methods and equipment. Engineers should also be encouraged to present papers and to serve on committees. Permission to attend seminars and conventions at company expense can also represent recognition for accomplishments.

Although professional registration is a personal matter, it bears directly on professionalism because of the recognition gained from belonging to an organization certified as professional and recognized by courts and governing bodies.

The Need for Educational Programs

Opportunities for self-improvement are particularly important to the development of professionalism. There are numerous ways that a company can help in this respect:

1. Reimburse engineers for the cost of under- and postgraduate tuition, fees, and supplies.
2. Allow some time off for educational pursuits.
3. Grant special awards, such as doctoral fellowships and scholarships, for outstanding performance.
4. Provide evening classes as part of degreed and nondegreed programs.
5. Present supervisory and management development programs on company time.
6. Bring authorities on specific fields to the plant to present seminars and exchange ideas.
7. Arrange special programs with universities when location makes pursuit of further education difficult.
8. Investigate new educational methods, such as closed-circuit television teaching.
9. Provide refresher courses in technical subjects and in communications.
10. Regularly compare company educational and professional development programs with those of competitors.

Offer Challenge: Foster Creativity

Challenge and creativity are vital to professionalism. A company should encourage creativity and boost technical competence by providing its engineers with interesting work and as much freedom as possible in solving problems. Although it is difficult to ensure that every engineer will always be working on a stimulating assignment, managers should be careful to distribute the routine work.

Adequate opportunities for promotion are critical to spurring technical achievement; however, a promotion must be considered in the light of each person's goals. Conferring administrative duties on an engineer who would consider them an unwelcome burden would be counterproductive if the job were accepted for the pay and stature. This problem has resulted in the well-known dual-ladder promotion plan, in which a company attempts to reward its technical people who prefer to remain technical on an equitable basis with those who enter into administration. Such a system must be genuine—not just exist on paper—if it is to succeed.

The Vital Role of the Supervisor

The performance of engineers is influenced most directly by the immediate supervisor. The relationship the supervisor establishes provides the professional climate in which performance either improves or deteriorates.

The supervisor will first want to make the work group effective by filling all positions with qualified people. They must then be provided with the tools, the information, and the atmosphere that promote performance.

When reviewing performance, the supervisor must consciously seek to establish a common meeting ground. This is best done through a sincere interest in employee welfare and progress, which can create the basis for a free and honest exchange of views on such things as the employee's strong and weak points, the quality and quantity of the employee's work, how the employee's value to the company can be raised, and the satisfaction the employee has derived from the job.

The supervisor should encourage the engineer to give his or her own opinion about his or her work performance and progress. This approach will enable the supervisor to better understand the engineer's attitudes and so be in a better position to make an evaluation.

Of course, professionalism requires that the engineer also recognize an obligation to acquire additional skills to upgrade capability and to use her or his talents in ways most beneficial to the individual and the company.

The Matter of Money

Despite surveys that purport to show that other factors are more important to job satisfaction, compensation is still an important element of professionalism. Undoubtedly, there are many instances when engineers do place salary second to the challenge of the job or a sense of accomplishment. Nevertheless, an adequate salary still remains the best method of rewarding employees. Although money alone does not guarantee job satisfaction, without this most powerful of rewards, other efforts to build a professional climate will fall short.

SETTING UP THE ENGINEERING TEAM

A top-notch engineering team is more than merely a staff of people hired to do engineering work. It is a carefully determined mixture of personnel having the variety of skills necessary to develop and execute engineering objectives. Managing such a team requires that the manager understand the functions of various members and get them to work together effectively.

The engineering team concept is gaining increasing favor as a means of stretching available engineering resources. By assigning the more repetitive and time-consuming tasks to specially trained support personnel, engineers can be freed to concentrate on the pure engineering portions of the design function.

Categorizing Team Members

Setting up the engineering team involves coordinating the roles of team members to achieve specified results. That coordination depends on knowing the specialized skills of the various types of engineering personnel and understanding the role that each type is best suited for.

Engineer. This person is primarily a creator of new products, processes, and systems, who develops ideas that can be used to solve specific practical problems economically. To accomplish this, he or she utilizes mathematics, physics, chemistry, and various other hard and soft sciences as tools.

Because of their specialized knowledge, engineers are qualified to "practice engineering." The "practice of engineering" refers to both creative design work and the consultation, investigation, evaluation, and planning that creative design entails. Engineers are thinkers as well as doers. Their most crucial contribution is their technical judgment. In addition, they are responsible for designing tests, evaluating data and plans, and inspecting construction for the purpose of assuring compliance with drawings and specifications.

Technologist. The technologist is typically a practical person who applies engineering principles and organizes people for the execution of industrial processes and methods. Ordinarily, the technologist oversees the construction and testing of the various components of a system that have been designed and developed by engineers. Technologists often become technical supervisors, filling a role that evolved during the 1960s and will continue to grow in the coming decades.

The technologist is a technically trained organizer and doer, rather than an innovator. As part of the engineering team, he or she may carry out a design function, but the design activity usually follows guidelines established by the engineer. In research and development, the technologist may serve as liaison between scientists and engineers on one hand, and craftspersons and technicians on the other.

The engineering technologist usually has a B.S. degree in engineering technology or has gained considerable technical experience on the job. The college education of a technologist is often an extension of a 2-year, full-time associate degree program to a four-year, full-time "bachelor of engineering technology" ABET-accredited program, which is usually offered by a college of engineering. The educational emphasis of the technologists' program is less theoretical and mathematical than that of their engineer counterparts, and it is more hardware- and process-oriented. There is less emphasis on courses in engineering analysis and design. There are forces afoot to make technologists eligible for professional licensure because their education can make them eligible after they have met the education and experience requirements of a state board of registration. This movement bears watching because it is a very sensitive subject.

Technician. The technician's role on the engineering team is basically one of support. This individual functions under the direction of engineers, scientists, technologists, or senior technicians. The technician's primary responsibility is to carry out proven techniques that are common knowledge among technical experts.

The technician often implements the ideas or carries out the "technical plans" of the engineer or scientist. She or he is usually a doer rather than an innovator or designer, although the technician may do some design work. Many opportunities arise to contribute to innovative changes in equipment.

The technician's education typically requires 2 years of full-time, college-level study or the equivalent in part-time study. The work is usually done in a technical institute or a junior or community college and leads, in most cases, to an associate degree. Graduates of these programs are being accepted in increasing numbers for transfer into bachelor of engineering technology (B.E.T.) programs with advanced standing. Technicians do not qualify for licensure as professional engineers on the basis of educational credit, but must go the long route of getting the experience that may allow them to meet the requirements of some board of registration. It should be noted that the chances for licensure are slim.

Where Does the Technologist Fit In?

Most employers and managers fail to understand the difference between engineers and technologists. Some of them do not really care, so long as people are made available for job openings that must be filled. The result is that engineering technologists are often hired to fill engineering-titled positions and vice versa. This practice is not only professionally and economically unsound, it creates expensive work force and human-relations problems in the workplace.

Some people claim that B.E.T.s are engineers, while others say they are technicians. The consensus seems to be that they are something between the two. Unfortunately, lines of demarcation have never been clearly drawn. In the professional literature, both technologists and technicians are usually called technicians. The issue is becoming increasingly important because the number of technologist graduates is growing every year and may continue to do so for the foreseeable future.

Technologists present a special "fit" problem. They must be integrated into the normal engineering routine in such a way that they fill important department needs and, at the same time, are able to pursue their own personal career objectives. Simply putting them into the usual draftsman or technician positions will not be enough. Their roles must be so designed that they become part of the normal promotional and advancement system.

The relationship between the engineer on the one hand, and the technologist and technician on the other, is virtually identical to that of the medical doctor and such supportive personnel as the medical technologist, nurse, and laboratory technicians. Use of the technologist and technician should be determined by the engineer, just as the doctor has the authority and responsibility to designate which medical tests will be performed by which personnel and how a patient will be cared for and treated.

Making Effective Use of Personnel

The job specifications covering engineers, technologists, and technicians should be carefully spelled out, because these specifications are all the manager has to go on when assembling a team. These job specifications should be detailed listings of tasks from the entry level, through the stage requiring minimal supervision, and up to the supervisor level. If the engineering-team concept is to be successful, there should be little or no overlap in these specifications.

The problem of utilizing technologists effectively can usually be solved by assigning them the work engineers prefer not to be bothered with. Of course this policy can be carried too far; there is a psychological dimension to consider. By giving technologists *certain* engineering responsibilities—and, at the same time, relieving engineers of time-consuming routine tasks—the motivation of both technologists and engineers can be improved.

In one department, for example, the personnel mix was changed from twenty-seven engineers and two technologists to seventeen engineers and twelve technologists. Staff morale and productivity soared—not because the people employed were more capable, but because the available skills and interests were more effectively utilized.

The technician has the same relationship to the technologist that the technologist has to the engineer. The technician must have a good background in the disciplines in which he or she will perform, but not to the extent of the technologist. The B.E.T. graduate must complete all of the courses taken by the associate-degreed technician, but in addition, the B.E.T. requires more advanced courses in math and computer hardware and software. Thus the technologist has a more well-rounded education than the technician and is more apt to advance into management positions. The technologist is not, however, qualified for high-level design work, which is the domain of the engineer.

The technician with an associate degree may be expected to perform any one or combination of the following tasks under the direction of a technologist or engineer:

- Drafting and detailing
- Routine trouble-shooting, computing, analysis, estimating, repairing, and inspection
- Preparing technical reports
- Reviewing and revising simple specifications
- Writing standard procedures and practices
- Carrying out elementary design
- Writing letters and directions for the engineer's signature
- Working as an engineering aide in a research laboratory

The engineer is the prime decision-maker in technical matters in an engineering department. Basically, the engineer judges the qualifications of technical specialists and the validity and applicability of their recommendations before such recommendations are incorporated into a design. Such decisions are usually made at the project level and higher. Examples of the kinds of decisions involved include:

- Selecting and evaluating engineering alternatives
- Selecting or developing design standards and methods
- Selecting materials to be used
- Selecting or developing methods of testing and evaluating materials or designs
- Reviewing and evaluating construction methods and controls, as well as evaluating test results, materials, and workmanship insofar as they affect the character and integrity of the completed work
- Developing and controlling operating and maintenance procedures

Satisfying Career Objectives

Engineers, technologists, and technicians may have different capabilities and responsibilities, but they have the same desire to be successful. In the engineering field, success means greater recognition, remuneration, and promotion. The engineering department and company organization must be structured so that all these personnel will be integrated into the normal promotional ladder.

This can be done by means of a "dual ladder" of progression: one for engineers and one for nonengineers. The engineer ladder has administrative, general practice, and specialist routes of advancement. The nonengineer ladder, for technologists and technicians, has administrative and general practice routes of advancement.

Entry on the engineer ladder can be accomplished in four ways: possession of a B.S. degree in engineering; registration as a professional engineer; being hired as an engineer on the basis of outside experience and education; and being certified on the basis of a successful completion of college-level studies, together with sufficient experience to qualify as having the equivalency of an engineering degree.

To illustrate the "equivalency" requirement, consider an individual who receives a B.S. degree in mechanical technology while working full-time. A study of his transcript shows that he has taken all of the major courses for a B.S. degree in computer technology. Further, the university granting his degree accepted him for graduate study toward a master's degree in computer technology. This evidence, supported by five years of experience in computer work and a good performance record, is used to justify his certification as having the equivalent of an engineering degree. This places him on the engineer ladder;* next step—senior engineer.

Nonengineer Ladder

Entry on the nonengineer ladder requires a B.E.T. degree for technologists, and an associate degree and two years of experience for technicians. Progression along the nonengineer ladder requires a demonstrated capability and a performance rating that is at least above adequate. The highest grades pose very stiff requirements, so that only the top performers can meet them. To avoid promotion on a seniority basis, promotions to the highest levels require top-management approval.

Use of the dual-ladder system, plus selected training programs, will help the supervisor to counsel nonengineer graduates about career possibilities. The ladder helps explain where everyone stands, what the possibilities are for the future, and what will be required to qualify for advancement. It also provides a way to absorb the nonengineer technical graduate and the nongraduate into an engineering organization with maximum benefit to both the individual and the organization.

Certifying Team Members

Presently, there are two forms of certification for members of the engineering team. One is the legal registration as an engineer by state examining boards, and the other is certification as an engineering technologist or technician by the Institute for the Certification of Engineering Technicians (ICET). The latter certification is not based on legal requirements; it is a recognition given by the engineering profession and industry. For information, write the National Institute for Certification in Engineering Technologies (NICET), 1420 King Street, Alexandria, VA 22314, or call (703) 684-2835.

At one time, a major weakness existed in the engineer registration procedures in the often casual manner in which engineering work experience was evaluated prior to granting permission to sit for the written examination; however, with the establishment of NCEE, and the cooperative efforts of the various state boards of registration, the evaluation of experience records has become more consistent.

---

* Two degrees plus experience would not necessarily imply that the individual has the equivalent of an engineering degree. One must determine whether the individual has the necessary math and science courses to qualify for equivalency.

THE ENGINEERING EDUCATION AND ENGINEERING TECHNOLOGY PICTURE—AN OVERVIEW

The manner in which some state boards of registration evaluate a B.E.T. graduate varies greatly from state to state.

Some state boards consider the B.E.T. as equivalent to the baccalaureate degree in engineering. Others give no educational credit whatsoever for the B.E.T. Most states grant a varying degree of credit toward the educational qualifications for the B.E.T. and require additional engineering experience before an individual can qualify to sit for the FE examination.

Some states may classify an engineering technology program as an engineering-related or science-related program and others do not. This issue is one that needs to be reconciled as soon as possible. Engineering educators can help by establishing some recognizable criteria with regard to Engineers' Registration that could be transmitted as recommendations to the various state legislatures and state boards of registration for implementation and incorporation into state law. In this way, the profession may be able to achieve a more uniform treatment by state boards of registration.

Based on my own on-hand experience working with engineers in private practice and industry, I highly recommended that our educators apprise their students of the differences between engineering education and engineering technology education, especially as it relates to their potential registration interests, so that they will not be either surprised or disappointed at some later point in their careers. As time passes, this condition will become more acute, since the enrollments in engineering technology programs are on the increase.

The student selecting an educational program to prepare for a career in the engineering field should be aware that, although many boards look for a B.S. engineering degree from an ABET-accredited program or board-approved program, many boards will evaluate transcripts of applicants who have B.E.T.s for maximum educational credit toward qualifications as a professional engineer. Interested persons should contact their boards for a copy of their state registration law and regulations regarding this important matter.

Passing the FE Examination

Engineering technologists often experience difficulty in passing the examination in the fundamentals of engineering (FE). Their passing percentage is low compared to those examinees who hold ABET-accredited degrees in engineering. A document prepared by NCEE[22] provides details designed to acquaint members of registration boards, professionals serving on society registration committees, and other professionals in industry and education with the structure and content of the FE examination and the performance of those candidates who sat for the examinations in October 1985 and April 1986. It is for their use only.

Differences in Curricula

The differences in the curricula selected by students in engineering technology and students in engineering may have to do primarily with the objectives. In the case of engineering technology, the curriculum may be more application-oriented, with less emphasis on higher mathematics, the basic sciences, and the engineering sciences. The professional engineering graduate program appears to be more strongly rooted in mathematics, the basic sciences, and engineering analysis and design, with the objective of a thorough education in creative engineering design.

The following summary is taken from the Candidate Performance Summary (referenced above) to provide some insight for the candidates interested in the Fundamentals of Engineering (FE) examination. Data cover the October 1985 (10/85) and April 1986 (4/86) examinations.

Fundamentals of Engineering Examination

Examinees	Date	Number of examinees	Number pass	Number fail	% Pass
All	10/85	15,698	9,445	6,253	60
	4/86	27,536	18,909	8,627	69
ABET code	10/85	12,402	8,298	4,140	67
1	4/86	22,979	17,074	5,905	74
ABET code	10/85	919	247	672	27
2	4/86	1,542	516	1,026	33
ABET code	10/85	688	341	347	50
3	4/86	935	538	397	58
ABET code	10/85	167	45	122	27
4	4/86	200	42	158	21
ABET code	10/85	1,522	514	1,008	34
5	4/86	1,880	739	1,141	39

Accredited by ABET: Code 1—4-yr or more engineering. Code 2—4-yr engineering technology

Not accredited by ABET: Code 3—4-yr or more engineering. Code 4—4-yr engineering technology. Code 5—None of the above.

SOURCE: Data courtesy of NCEE.

SUMMARY OF STATE REGISTRATION LAWS

Engineers simply cannot accept state or regional boundaries as limits on their activities, because of the nomadic nature of engineering practice itself. The next job, the next project, may literally take the engineer anywhere. It therefore appears incumbent upon every engineer to know the registration law pertaining to her or his activity and to know exactly what are the legal requirements and provisions of the law which must be met for professional practice.

Only the states can enact licensing statutes based on the police power, but these statutes are by no means uniform from state to state. They have no extraterritorial application. Thus, an engineer registered in one state does not necessarily have the legal right to practice engineering in another state. The second state has the legal right to refuse registration to the engineer if that person is not able to meet its qualifications. This is true not only in engineering but also in law and medicine. States' rights, a basic structure of our federal government, are involved, and the situation cannot be cured overnight. Only through the gradual adoption of uniform registration laws will it be improved.

There is a question as to whether or not full reciprocity between states in matters of registration of engineers will ever materialize, but the chances are better for engineering than for either law or medicine. This may be because engineering practice is nomadic and much less localized in nature than any of the other professions.

MODEL LAW

Almost as soon as Engineers' Registration made its appearance, there arose a movement to seek uniformity in the laws. A national registration law was suggested by some, but because of state prerogatives and local conditions, it became obvious that such a law was not feasible.

Requirements for registration are fundamental, and most states adhere rather closely to the so-called "Model Law," in some form and to some degree. As we learned, registration is basically and legally for the protection of the public, and all state registration laws have a provision to this effect. The wording in most of the statutes is similar to that of the Model Law, and probably the most significant is the following, quoted from the act of Congress of the United States regulating the practice of engineering in the District of Columbia:

> In order to safeguard life, health and property and promote the public welfare, the practice of engineering . . . is hereby declared to be subject to regulation in the public interest. It is further declared to be a matter of public interest and concern that the profession of engineering merit and receive the confidence of the public and that all qualified persons be permitted to engage in the practice of engineering. All provisions of this Act relating to the practice of engineering shall be construed in accordance with this declaration of policy. Any person engaged in or offering to practice engineering . . . shall submit evidence that he is qualified to practice and shall be registered.

It took the civil engineers to set the pattern for the Model Law as it stands today. They defined engineering and set up the machinery for regulation. The first draft of the law appeared in the *Proceedings of the American Society of Civil Engineers* for January 1911. And to the present day the society has played an important part, along with other engineering groups, in its development.

Revisions in the Model Law have been many over the ensuing years. In March 1943, there was a revision made and a new classification added, namely, Engineer-in-Training. This was used to earmark young persons as prospective professional engineers.

By means of the Model Law and the state registration laws, which are its reflective counterparts in many respects, the legislators and the public have been made to realize that engineering is a learned profession. Educational requirements for the licensing of engineers have been written into the laws. Demonstration of good moral character, evidence of completion of academic and professional education, and evidence of field experience of a character satisfactory to the state boards have also been included.

For copies of the Model Law (1984 Rev.), interested parties are encouraged to write NCEE.

THE NCEE MODEL LAW GUIDE—A SUMMARY FOR ENGINEERING

The intent of NCEE in preparing the *Model Law Guide* is to present to the states a sound and realistic guide which will provide greater uniformity of qualifications for registration, which will raise these qualifications to a higher level of accomplishment, and which will simplify the interstate registration of engineers.

The primary purpose of NCEE is to serve as an organization through which its member state and jurisdictional boards can counsel and act together to better discharge their duties as individual, autonomous regulatory agencies dedicated to the protection of the public life, health, and property. Standards presented here have been approved by the NCEE member state regulatory boards and represent optimum, realistic levels of qualifications for initial and subsequent registration to ensure protection of the public interest.

As revised through 1984, this guide contains three chapters designed to assist legislative counsels, legislators, and NCEE members in preparing new or amendatory legislation. There are chapters which deal with the registration of engineers and the duties of the responsible board.

Registration of Engineers

Section 1 deals with general provisions while Section 2 deals with definitions.

Engineer
Professional Engineer
Engineer-in-Training
Practice of Engineering
Consulting Engineer
Board of Registration
Responsible Charge

Section 3. Board Appointments, Terms
Section 4. Board Qualifications
Section 5. Board Compensation, Expenses
Section 6. Board—Removal of Members, Vacancies
Section 7. Board Organization and Meetings
Section 8. Board Powers
Section 9. Receipts and Disbursements
Section 10. Records and Reports
Section 11. Roster

Section 12. General Requirements for Registration
  As a Professional Engineer
    Registration by Comity or Endorsement
    Graduation, Experience, and Examination
    Engineering Teaching
  As an Engineer-in-Training
    Graduation and Examination
Section 13. Application and Registration Fees
Section 14. Examinations
  Engineering Fundamentals
  Principles and Practice of Engineering
Section 15. Certificates, Seals
Section 16. Expirations and Renewals
Section 17. Reissuance of Certificates
Section 18. Public Works
Section 19. Disciplinary Action—Revocation, Suspension, Fine, or Reprimand
Section 20. Disciplinary Action—Procedures
Section 21. Violations and Penalities
Section 22. Authorization Certificates
Section 23. Exemption Clause
Section 24. Invalid Sections
Section 25. Repeal of Conflicting Legislation
Section 26. Effective Date

The *Model Law Guide* also covers land surveying engineers and land surveyors.

The above was taken from the NCEE publication pamphlet *Model Law—A guide prepared by the National Council of Engineering Examiners for use by its Member Boards and the State Legislatures in the interest of promoting uniform laws for the registration of Engineers and Land Surveyors*. 1984 Revision.

STATE REGISTRATION LAWS

Certain provisions characterize most of the registration laws. One of the most important is that only those who have satisfied the requirements prescribed by law may be registered as professional engineers. Another is that the state legislature itself has the responsibility of determining who is qualified to practice and who should be rejected as not qualified. Thus, for the protection of the public, the exclusive right to practice in the field of engineering has been reserved for those who satisfy the legally prescribed qualification requirements. The state legislature itself has the task of determining what the exact requirements are and when they have been satisfied. This practice follows an age-old custom whereby the qualifications of learned individuals are judged only by those of similar learning, who have already established their qualifications.

The profession is expected to administer itself, and the laws give it the right of self-preparation and self-determination.

The board of examiners, in some states known as the board of registration, is the official committee of the profession. It is usually appointed by the governor. Board members must at least meet the minimum requirements to practice under the law, but the persons appointed are usually well-known practitioners in the state. The board, through the state legislature, determines what constitutes adequate education and experience and authorizes an applicant to practice only when that person has been found to have satisfied the full qualification requirements.

It might be asked, "How was the first board appointed?" For example, when the bill which Governor Alfred E. Smith signed into law in New York State provided for a board of examiners and land surveyors, there just weren't any people to fill the requirements, for the law recognized only those registered in New York. The AIEE, ASME, and ASCE made nominations, but no board could be appointed. Finally, the Iowa board agreed to come to New York and to look over the candidates. The trip was made, and five men were found qualified and recommended for licensure.

Enforcement of the registration laws falls upon the state legislature and the state's attorney's office. When Engineers' Registration was in its infancy, many disregarded it or were ignorant of it because of lack of enforcement procedures and lack of publicity. Since then, the machinery of enforcement has been established, and people have been arrested, tried, and fined for practicing engineering before the public without a license.

Each state has its own statutory qualification requirements. Its application forms are designed around them and include whatever additional information the board of examiners believes necessary for consideration; however, the main core of the application for each state is similar. In addition, the applicant must attest to the veracity of all statements contained in the completed form.

Amendments to registration laws are made from time to time as a result of changes made by state legislatures which may reflect changes in practice, upgrading of statutes, or other reasons. Some aspects of the laws are often challenged as to their constitutionality. Attempts to have the laws declared unconstitutional have been many. In one instance, the plaintiff's contention was: "Conceding that such regulation is within the police power of the state, the statutes setting up the modus operandi for the (board) to set up standards of qualification and to conduct examinations constitutes a delegation of legislative powers to a committee and is, therefore, unconstitutional." The Supreme Court found no merit in the attack upon the statutes and affirmed the judgment of the trial court.

SUBJECTS COVERED IN THE LAWS

Subjects covered in the various state registration laws are as follows:

1. Application Forms for Registration
2. Duty of Attorney General and Other Public Officers
3. Bond, When Required of a Professional Engineer
4. Charges, Hearings, Findings, Review, and Appeals
5. Constitutionality of Registration Law
6. Corporations and Partnerships
7. Contracts in Violation of Law, Void
8. Credits, Education, Experience
9. Definitions
10. Engineer-in-Training, Enrollment
11. Exemptions
12. Expiration and Renewal of Registration
13. Fees, Registration, Renewals, Etc.
14. Injunction Restraining Unlawful Practice
15. International Registration under Treaties
16. Legislative Authority
17. Liens Securing Payment for Engineering Services
18. Offical Title, Address, and Duties of Board
19. Overlapping Professional Practice
20. Powers of the Board, Rules, Regulations
21. Public Works, Registered Engineering Services Required
22. Purpose of Registration Law
23. Qualifications of Registration
24. Reciprocity, Comity, Registration of Nonresidents
25. Registration, Property Right
26. Registration, Certificate of

27. Registration in Fields and Branches of Engineering
28. Replacement of Lost, Destroyed, or Mutilated Certificates
29. Revocation, Suspension, and Reinstatement
30. Roster of Professional Engineers
31. Seal and Signature of Registered Professional Engineer
32. Temporary Permit, Special Permit for Single Project
33. Violation, Punishment

See the pertinent state laws for detailed information about any of these categories.

EXEMPTIONS

In general, exemptions from registration fall into the sixteen categories listed below:

1. Nonresidents, who may practice no more than 30 days.
2. Nonresidents or recent arrivals in a state, if they have filed an application for registration, who may practice more than 30, 60, or 90 days in states with these minimum requirements, until action is taken on their applications by the state board of examiners or any other body having jurisdiction.
3. Employees of registrants or of persons exempted from registration, if practice does not include responsible charge.
4. Officers and employees of the United States government (not in all states).
5. Elective officers or employees of the state, county, etc.
6. Employees of transportation companies, or of corporations engaged in interstate commerce, or of public utilities, subject to regulation by Public Service Commission.
7. Consulting associates of registrants, if nonresidents and qualified to practice in own states.
8. Practitioners of other legally recognized professions.
9. Services performed by locomotive, stationary, or power-plant engineers, etc., or in operation of machinery or equipment.
10. Assistants to or employees of registered engineers.
11. Those with overlapping practice, i.e., engineering and architecture.
12. Contractors, superintendents, or foremen for execution of work designed by professional engineers.

13. Individuals, firms, or corporations who practice on property owned by such individual, firm, or corporation, if not involving health, safety, and well-being of the general public.
14. Employees of manufacturing companies performing services.
15. Those engaged in construction of private dwelling or building.
16. Miscellaneous. See specific state law exemption provision.

It should be noted that many persons concerned with licensing now favor registration of both principals and all engineering employees of firms (partnerships and corporations) engaged in any work of an engineering nature where health, life, safety, and social well-being of the general public are involved. And today almost any design, operation, or process undertaken by an engineer has public implications—from buildings, power plants, and bridges to electric motors and appliances. State registration laws are being amended to attempt to encompass all eventualities.

It is accepted practice that a person registered in one state and wishing to practice in another request permission, in writing, from the registration board of the second state. The board will then authorize the engineer to proceed, under special permit or an exemption clause of the registration law. One must not under any circumstances proceed without such authorization. In this way, the nonresident registrant may save great loss and embarrassment. In every case, the board has the last word.

ENFORCEMENT

In most states, the attorney general's office is charged with the duty of enforcing the provisions of the registration act. They make investigations of violations and present the facts of such violations to either the attorney general or the district attorney for prosecution. Individual licensed professional engineers can be of great help as a policing power by reporting violations immediately to the secretary of the board of examiners. The *Registration Bulletin*, published by NCEE, gives state law enforcement activities for a number of states. The following tabulation lists the various ways the law may be enforced:

- Investigation and settlement occur out of court if possible.
- Board secretary and board members investigate and case is referred to the attorney general.
- Society committee investigates; board sues violator if warranted.
- Violations are reported to law enforcement officers.

- Law is enforced by all constituted officers of law.
- Settlement is made on friendly basis unless there is willful violation.
- Board attorney cooperates with state attorney general.
- Investigation is made on receipt of written complaint and handled through office of attorney general.
- Charges are filed and hearing is conducted in three months. Unanimous vote by board members is required to convict and/or revoke license.
- Complaints are investigated by board's representative and then filed with county prosecuting attorney.
- Violators may be restrained by injunction.
- Violation considered misdemeanors are punishable by fine or imprisonment.
- Letters are written to violators; then case is submitted to attorney general.
- Charges are filed and then hearing is held.
- Investigation by inspector occurs, followed by prosecution by attorney general and prosecuting attorneys.
- Disciplinary action is made against registrant by the board; criminal action is exclusively at the discretion of the prosecuting attorney.
- Civil authorities and board are responsible for enforcement.
- Complaints are filed with county solicitor, who prosecutes. Board, after hearing, may revoke, suspend, or annul professional's license.
- Violator is given two warnings, then case is given to district attorney.
- Criminal violations are taken to court and all disciplinary cases are presented to board.
- Board enforces, with advice of attorney general; there is cooperation with state societies committee.
- Secretary of justice is notified; suit is taken to superior court.
- Case is referred to assistant district attorney who represents board.
- Engineering societies investigate and settle.

A public uneducated in this field is the greatest obstacle confronting enforcement. It is essential to establish a public-relations program to inform the public that there are registration laws and that all practicing engineers must be licensed.

WHAT GOOD IS A PUBLIC MEMBER ON A REGISTRATION BOARD?

Public members have been included on boards of registration in the various states. I would like to focus on ways in which public members can become valuable assets in helping registration boards discharge their responsibilities as they relate to the welfare of the public.

The natural first reaction of the professional board members was that a nonprofessional, being unqualified to judge the technical competence of an applicant for registration, was at best a useless addition to the board, unable to contribute significantly, and at worst was a potential burr under the saddle blanket. These have proved to be only fears, not unavoidable consequences. Most public members have proved to be assets.

Let's look at the facts. Public members are now on about 60 percent of NCEE member boards. They are usually consumer advocates selected for their good judgment and constructive ideas; but they came on registration boards ignorant in large part of what they are supposed to do, and with no prior knowledge of how the board operates or is performing its duties. "I went on the ______ board all ready to fight. I was sure I'd have to change everything," said one nonprofessional board member recently, "but I feel a willingness [on their part] to protect the public. They want to get rid of the bad guys, but they don't have enough resources. There are not enough investigators, and the legislature just keeps increasing our authority and cutting our budget." Perhaps a public member, adequately informed, may sometimes be a more effective advocate before the legislature than a professional member.

The key is to become adequately informed. A public member has the responsibility of knowing the law, the board rules and regulations, and the reasoning behind board actions. It has been suggested that the board president can promote a feeling of productive participation by assigning suitable specific duties. In one instance, the public member was unanimously elected board secretary.

The end purpose of a registration board is to promote the public welfare by screening out incompetent applicants and weeding out misbehaving practitioners. The role of a public member is to assure the citizenry that the board is trying to do these things and is not just looking out for the interests of its professionals. This role requires a public member to be alert to all of the facts underlying board actions. Although a board may not be guilty of favoring its professionals by purposely limiting competition, such as denying registration by comity or unlawfully limiting the granting of temporary permits to qualified applicants, it still may be falling short of discharging its duty by its inaction, or inadequate action, in cases of alleged misconduct of practitioners. This is almost always caused by lack of funds or lack of qualified personnel. Who can be in a better position than a public member to advertise the board's requirements for fulfilling its mission? This is an important area for a public member to work in, and one that is often overlooked.

Time has borne out the fact that public members on boards are a good thing for the engineering profession in terms of showing that there is nothing secret or sinister in the way boards handle their business. Also such persons have proven to be good allies in helping the public to understand the true purpose and public benefit of the registration laws.

In a real sense, public members are the public relations officers of their registration boards and have the important responsibility of keeping the legislature and the public aware of what is needed to assist their boards in effective law enforcement for the protection of the public. The following table lists the number of public members by state in 1985:

Number of Public Members on State Boards of Registration

State	No.	State	No.	State	No.
Alabama	0	Kentucky	1	Ohio	0
Alaska	1	Louisiana	0	Oklahoma	1
Arizona	0	Maine	1	Oregon	2
Arkansas	2	Maryland	2	Pennsylvania	3
California	7	Massachusetts	0	Puerto Rico	0
Colorado	2	Michigan	2	Rhode Island	0
Connecticut	3	Minnesota	5	South Carolina	1
Delaware	1	Mississippi	0	South Dakota	1
District of Columbia	0	Missouri	1	Tennessee	1
Florida	2	Montana	2	Texas	3
Georgia	1	Nebraska	0	Utah	1
Guam	0	Nevada	0	Vermont	1
Hawaii	3	New Hampshire	1	Virginia	0
Idaho	0	New Jersey	2	Virgin Islands	0
Illinois	0	New Mexico	1	Washington	0
Indiana	1	New York	2	West Virginia	0
Iowa	2	North Carolina	2	Wisconsin	4
Kansas	1	North Dakota	0	Wyoming	0
		Northern Mariana Islands	2		

5

ENGINEER-IN-TRAINING PROGRAM

The registration laws of all states now provide for the enrollment or certification of engineers-in-training (EIT). Engineer-in-training programs are being successfully administered in all states, and EIT examinations are conducted periodically.

In 1943, the Engineer-in-Training program was established and included in the Model Law. By 1949, a considerable amount of experience had been gained in its administration, and it was reported that at that time some nineteen state boards had an EIT program in operation and had certified a total of over 11,000 EITs.

The time may be coming when all registrants will have been, at one time, enrolled as engineers-in-training, and perhaps some day these enrollees will have to be graduates of ABET-accredited engineering programs. For copies of the pamphlet *Accredited Programs Leading to Degrees in Engineering* (1985), write the Accreditation Board for Engineering, 345 East 47th Street, New York, NY 10017.

The hope is expressed in some quarters that the time will come when all professors who teach engineering courses will themselves be registered professional engineers. And there are some engineering schools and colleges where this is now true. In the future, the engineering student, the engineer-in-training-to-be, will have a better understanding of engineering as a profession.

THE ENGINEER IN TRAINING

The engineer-in-training movement has become the gateway through which young engineers will later enter the profession as registered professional engineers. This path to the professional engineer's license has been provided for, as we saw, in the licensing statutes of most states.

Engineering students have pretty well made up their minds that engineering is the career for them. The engineering faculty does not have to "sell" engineering to them. They are already convinced.

However, the profession of engineering in its more abstract applications does require some presentation by the faculty to engineering students. To teach students to think like professionals and to act like professionals presents a challenge. The importance of the professional attitude is somewhat more difficult to put across than technical subject matter.

Formerly, faculty members were satisfied to emphasize the importance of an individual code of ethics, but they have come to realize that this in itself is not enough. It has to be complemented by professional training.

The most important tool the engineerng educator possesses is the inspiration he or she can offer students. The degree to which a student will develop depends on this inspiration and guidance.

Target the engineer-in-training certificate, and the engineering professor and student will automatically realize the importance of professionalism. Supplementing technical training with professional training will help make the college senior eager to become an integral part of the engineering profession, when the student leaves the protective atmosphere of the campus and enters practice to gain technical experience. The student who aims at the EIT certificate has then completed three of the four steps toward professionalism—matriculation, graduation, enrollment as an engineer-in-training, and finally registration as a full professional engineer.

The major objectives of the engineer-in-training program have been stated in Chapter 1. They are the means for welding a link between the student and the professional engineer. The once-serious gap in the path to final licensure has been closed, with resultant greater control by the engineering profession over its membership.

By the same token, the general public also receives increased protection, from the very fact that the engineering profession is in a more responsible position with regard to preparation and training of its members. What is good for the public is good for the engineer-in-training, in increased public respect for the profession as a whole.

MODEL PROGRAM

In view of the early experience gained in the operation of the EIT program, a "model" program was developed. It was instituted to increase uniformity of law and to widen reciprocity among the various states, in an attempt to ease the EIT's transfer of application from one state to another.

The factors to be considered were:

1. Greater publicity will provide greater effectiveness. The EIT program should be brought to the attention of as many as possible potential candidates for registration as professional engineers. Contact on campuses before graduation is most effective for this.
2. Participation in this on-the-campus program should be as inexpensive as possible.
3. By bringing the program to the student and making it convenient and palatable, a higher participation rate can be expected.
4. Since engineering is nomadic by nature and since the ultimate goal is registration in the engineer's home state, the participant must be provided with the protection of a high degree of comity among states.
5. Since a large number of state boards now require a written examination for registration, and since all states require sampling of some sort, *a written examination on theory and fundamentals is essential if some degree of comity is to be eventually achieved.*

PROGRAMMING FOR RECIPROCITY FOR EIT

In order to promote a high degree of uniformity in standards and to enhance the possibility of improved reciprocity relations between the various states, the following program has been recommended as a guide to the state boards by NCEE.

1. *Written examinations required.* Enrollment as an engineer-in-training is presently based upon a written examination given to graduates of and seniors about to graduate from approved engineering curricula and to nongraduates who in the studied opinion of the state boards involved have had experience and/or education equivalent to that obtained through completion of a standard four-year engineering curriculum.

2. *Type of examination.* A written examination is now called for by all state boards, covering one day of eight hours. Part of this time is devoted to the common basic subject matter of all engineering curricula. See Chapter 10 for NCEE examination coverage.

Emphasis is placed on fundamentals, and questions are arranged to test the student's analytical ability rather than his or her memory. It is common practice in an open-book examination to time the examination so that the student (candidate) may use books as reference material but will not have time to study during the exam. There is now a sufficient choice of questions on each exam to provide a graduate of any first-class engineering school with a reasonble set of questions to answer.

3. *Time and place of examination.* Where conditions permit, examinations are held on the campuses of accredited engineering schools within the state, as well as in other predesignated locations in the state. The dates for the examination are fixed throughout the country.

4. *Publicity.* As a preliminary to the FE examination, the students should be made acquainted with their responsibilities as members of the engineering profession and with the requirements and procedures for ultimate registration. This information is usually presented by members of the board of examiners, by prominent registered engineers, or by members of the faculty who are registered professional engineers. This should be an educational program and *not* a selling program.

5. *Fees.* Fees are nominal but sufficient to cover the cost of administering the program.

INTERSTATE RECOGNITION OF ENGINEERS IN TRAINING

The problem of recognition of an EIT certificate issued in another state to an applicant for registration as a professional engineer has become increasingly important. Registration boards are receiving numerous requests and inquiries from persons to whom they have issued EIT certificates who are currently living in other states. It has been the practice of many registration boards to insist that applicants be registered in the state of their legal residence before permitting them to become registered in a state of which they are not residents. However, in some states this policy is not rigorously followed, especially where these engineers are daily commuters living in an adjacent state. A case in point is New York City and adjoining cities in New Jersey and Connecticut.

Most states now accept the EIT certificate if the individual holding it has an accredited degree and has passed the FE exam.

Our various state boards are emphasizing more and more the desirability of issuing EIT certificates only to those who have graduated from an ABET-accredited program in engineering. They also recommend, in each and every case, the passing of a written examination of 8 hours in length in the NCEE format, i.e., the FE exam.

As mentioned above, the person holding certification as an EIT should be encouraged to complete final registration in the state of legal residence. Exceptions were stated previously. Another exception might possibly be made in the case of a person residing in a state where a final written examination is not required for registration (which would be rare). It might be to the advantage of the prospective registrant, under such circumstances, to return to the state from which the EIT certificate was obtained and complete registration by examination in that state. In that way, the person might, at a later date, become eligible for *registration by endorsement* in other states requiring examinations of all applicants for registration.

THE FE EXAMINATION

The FE written examination, also known as the preliminary or first day's examination, tries to test the candidate's facility in mathematics and engineering theory. The boards want to know whether or not the candidate understands basic engineering principles. Naturally, the test cannot cover every detail of engineering know-how, for it is merely a sampling process. The objective of any sampling process is to reach sound conclusions with a minimum of time and effort. It has been found that many small samples, well distributed, will give a more accurate indication of quality than can be obtained from a few large samples. Good distribution is essential for the elimination of one-sidedness in examinations.

The curricula of college courses in the various branches of engineering differ in many details, but there is a core of common material. This is the material that is basic to engineering and is covered in the FE examination. See Chapter 10 for details about exam material.

In addition to the above, many engineering schools include ethics and business law or contracts in their curricula. These have been found to be important subjects and are beginning to appear in the examination material for EIT, although they are more effectively included in the professional, or second day's, examination for full licensure, after the EIT has gained some experience in the workplace.

It is generally agreed among the state boards that the fundamentals include mathematics through the integral calculus, one year each of chemistry, physics, and mechanics (including statics, kinetics, and strength of materials), and at least two of the following courses; hydraulics or fluid mechanics, electrical engineering, and thermodynamics or heat power.

Mathematics

The basic importance of mathematics becomes apparent from an inspection of license examinations. The physical sciences and engineering principles rest upon a foundation of mathematics and involve the practical application of mathematics. The ease with which mathematical problems can be devised, together with the importance of the subject matter, have led to their common use in license examinations. The EIT candidate must pass a written examination which tests skills in mathematical manipulations as well as skill in applying answers to the solution of engineering problems.

Open-Book Examination

All states administer the NCEE examination, which is open-book for the Fundamentals of Engineering and the Principles and Practice of Engineering.

The open-book examination can be made effective by careful wording, subject coverage, and careful problem selection. Good performance on such a test requires practical insight into terminology. Problems involving basic concepts will be included. Questions regarding judgment are usually relegated to the PE examination for the more seasoned candidate.

Grading

The FE examination is now machine-graded and the examinee has the choice of selecting the method of solution, either in the English system or the metric system.

WHAT THE EIT CERTIFICATE IS NOT

There has been some evidence of false representation of what the EIT certificate is. Apparently, it has not always been made clear to the applicants for certification (enrollment) that they are being granted the privilege of taking the examination on engineering fundamentals at a time when it is most convenient for them to do so but that they have not been given the privilege of conducting themselves as registered professional engineers.

The boards report that a considerable number of EITs have been applying for final registration within 2 or 3 years after graduation, without waiting for the statutory 4 years' experience (as of application date) which is required as a minimum in most states. Many of these persons have attempted to include summer vacation experience, subprofessional experience gained prior to college entrance, experience after 1 or 2 years of college education, or military experience involving nonengineering research or development as satisfying the requirement of 4 years of approved engineering experience.

While such persons are to be commended for their expression of zeal and ambition to become registered as quickly as possible, they should have been informed at the time of application for EIT certification that most registration boards require them to have, subsequent to graduation, a minimum of 4 years of acceptable experience (as of application date) before being eligible for final registration.

TIME LIMITATION ON EIT CERTIFICATES

The validity of the EIT certificate varies so widely from state to state that interested persons should write to their state board for the latest information on this subject.

6

REQUIREMENTS FOR REGISTRATION

The requirements for registration cannot be precisely defined, because of differences in the registration laws of the individual states and territorial possessions; however, there are seven general requirements, the details of which may vary to some degree from state to state:

1. *Age.* Minimum age for full professional engineer's license is twenty-five. For the Engineer-in-Training (EIT) certificate, the applicant must be twenty-one.
2. *Citizenship.* All states do not require United States citizenship.
3. *Graduation.* Applicant must hold a high school diploma or the approved equivalent.
4. *Degree.* The standard requirement is an undergraduate degree in engineering or related science from an ABET-accredited program, or a board-approved curriculum.
5. *Experience.* Evidence of experience sufficient and "qualifying" as of the date of application must be presented.
6. *Character.* References attesting to the applicant's good moral character and integrity are needed (usually five in all, three of whom must be licensed professional engineers).
7. *Examination.* Unless waived by the board, the applicant must pass all parts of the professional engineering examination administered by the board.

UNIFORMITY IN REQUIREMENTS SOUGHT

Evidence of progress in achieving uniformity in procedures for administering the registration laws is noteworthy. The possibility of achieving completely identical laws may be remote, but all boards agree with and try to act upon the principle that a determination of competency is the proper objective.

Working toward uniformity, the Committee on Uniform Laws and Procedures of NCEE set forth a list of requirements as a goal for registration to be adopted by state boards, although many have not used these criteria as their sole basis for decision:

1. Register engineers only.
2. Register as "professional engineer."
3. Do not register in branches of engineering, but list in roster "field of training and/or experience."
4. Use as minimum requirements for registration as a professional engineer: graduation from a school having a curriculum approved by the board, plus 4 years of acceptable experience, plus 16 hours of written examination. (Registration of applicant with outstanding record of engineering education and practice is entirely up to the board.) It should be noted that many state boards still use additional criteria for nongraduates, and they do not limit registration to graduates of schools offering only ABET-accredited programs in engineering.

Minimum requirements for engineers-in-training: graduation from a school having a curriculum approved by the board, plus 8 hours of written examinations.

ARE YOU READY FOR ENGINEERS' REGISTRATION?

Whether you are an unlicensed engineer, a recent engineering graduate, or a student, the following compendium and checklist brings you the more pertinent facts that can get you started on the road to professional licensure and open up for you the greater opportunities available to the licensed engineer. Also included is a sample letter of request for application materials. Fill in the appropriate boxes and compare your answers with the answers that follow.

1. Age: Years ( )
2. U.S. citizenship: Yes ( ) No ( )

3. Presently employed in engineering work? Yes ( ) No ( )
4. Can you obtain the signatures of three licensed professional engineers, persons to whom you reported or with whom you have had professional associations? Yes ( ) No ( )
5. High school graduation or approved equivalent? Yes ( ) No ( )
6. ABET-accredited degree in engineering? Yes ( ) No ( )
7. ABET-accredited B.E.T. degree? Yes ( ) No ( )
8. Science degree? Yes ( ) No ( )
9. Accredited degree in engineering plus ( ) years of engineering experience
10. Science degree plus ( ) years of engineering experience
11. B.E.T. accredited degree plus ( ) years of engineering experience
12. Engineering experience in armed services? Yes ( ) No ( )
13. Engineering experience in construction contracting? Yes ( ) No ( )
14. Engineering teaching experience? Yes ( ) No ( )
15. Master's degree in engineering? Yes ( ) No ( )

Now compare these answers with your answers.

1. Age twenty-one is legal minimum for Engineer-in-Training: twenty-five for PE. (In New York 19 for EIT; 21 for PE)
2. U.S. citizenship no longer required.
3. Answer must be "yes" to meet minimum statutory requirements.
4. Boards may require signatures of at least three licensed professional engineers. Answer must be "yes." New York requires signatures from New York–licensed professional engineers.
5. Answer must be "yes."
6. Answer must be "yes."
7. With ABET-accredited B.E.T. degree, may need 8 years of approved engineering experience or more to qualify. Transcript of record is usually evaluated by board.
8. Science degree plus master's in engineering from an ABET-accredited program has three-fourths equivalency value.
9. Graduation from an ABET-accredited engineering program may qualify for FE examination immediately after graduation. Then, with 4 years' approved engineering experience, may qualify for PE exam. In most states students who are progressing normally toward graduation in an engineering program will be admitted to the FE exam prior to graduation.

10. See item 8 above. Experience record will be evaluated on basis of certified transcript of college record and experience.
11. Time can vary from state to state. Check with your board.
12. If answer is "yes," experience must be engineering in nature, character, and extent. Write up experience record to reflect this.
13. If answer is "yes," proceed carefully. Basic principle underlying the practice of professional engineering is "design." Most contracting firms in the field of contruction execute designs and specifications formulated and worked out by engineers. While contracting work may provide the individual employee with a knowledge of construction methods, it will not, per se, increase that person's knowledge of design to a degree that will justify qualification.
14. Must be involved in an accredited ABET engineering school. Boards will give credit if the applicant is in responsible charge of engineering teaching, has experience up to 3 years, is at college level and satisfactory to the board, or is teaching engineering subjects and has the rank of assistant professor.
15. A master's degree in engineering from an ABET-accredited program of study during the daytime (i.e., full-time) is equal to 1 year accredited experience; part-time study (in the evenings) does not count toward accredited experience.

Since boards retain the right to change their rules and thinking, it is suggested that the interested engineer contact her or his board for any clarifications. All applicants must meet the statutory requirements; the board's decision is final.

7

WHAT IS QUALIFYING EXPERIENCE?

If you have the experience and professional know-how, your registration board wants you to have your professional engineer's license. It isn't their job to throw roadblocks in your path. Actually, the opposite is true. Boards want to advance recognition of professional engineering by awarding licenses to the greatest possible number of qualified engineers.

Engineering experience submitted by all applicants for licensure must be of a character and grade satisfactory to the board of examiners. It must at all times be remembered that the number of calendar years of experience does not necessarily equal the number of years of board-approved experience. Although acceptable experience is something defying exact definition, all boards insist that "approved" experience be broad in scope and of such nature as to have developed and matured the applicant's knowledge and judgment. The boards have the final say-so as to what constitutes approved experience.

Educational background is considered here only insofar as it affects the legal minimum period of qualifying experience. *Examinations, as a requisite for registration, are considered to be separate and distinct from qualifying experience.*

As we previously learned, the basic objective of registration is to assure minimum competence and sound judgment in professional engineering for reasons of safety and health of the public and protection of property. The basic objective of the educational requirements prescribed or given credit in the various state laws is to assure the public that the applicant has the requisite knowledge of mathematics and engineering sciences to enable him or her to use them intelligently in the solution of engineering problems.

The basic objective of requirements of qualifying experience is to make sure that the applicant has acquired, through actual engineering practice that is of suitable character and caliber, the professional judgment, capacity, and competence in the application of the engineering sciences requisite to registration.

The kind of experience that will be approved should be acquired as much as possible within a single 4-year period, in order to develop fully competent professional stature. But for acceptance and registration at the bare end of the minimum prescribed period, the kind of experience acquired would probably have to be above the average.

The quality of experience should be such as will demonstrate that the applicant has developed technical skill and initiative in the correct application of engineering science, sound engineering judgment in the creative application of engineering principles and in the review of such application by others, and the capacity to accept responsibility for engineering work of a professional character.

ACCEPTABLE EXPERIENCE

When an applicant files for the professional engineer's license, the question of what constitutes acceptable experience often is given only passing attention by the applicant. As a result, this important information is not well documented in the application.

An acceptable amount of experience is often found to require more than the time elapsed since the applicant's graduation. As we saw, state boards require a minimum of 4 years of qualifying experience. And quite often this lack of the right experience proves a big disappointment to those who are qualified in all other respects.

NCEE defines qualifying experience as

> the legal minimum number of years of *creative* engineering work requiring the application of the engineering sciences to the investigation, planning, design and construction of engineering works. It is not merely the laying out of details of designs, nor the mere performance of engineering calculations, writing specifications or making tests. It is rather a *combination* of these things plus the exercise of sound judgment, taking into account economic and social factors, in arriving at descisions and giving advice to the client or employer, the soundness of which has been demonstrated in actual practice.

For the newly graduated engineer, some routine work is unavoidable, and it may even be desirable. But to obtain the proper amount of qualifying experience the engineer should seek the widest possible responsibility. Boards value experience of a broad, diversified character more highly than largely specialized activities.

APPLICATION OF IMPORTANT PRINCIPLES

Knowledge of the basic fundamentals of science and mathematics alone does not fully satisfy the requirements. Evidence of ability to use these fundamentals in the design of engineering works is required. Design in its broadest sense, not solely confined to making drawings and computations, is the ultimate intent.

Problems in, for example, mechanical engineering cannot be handled adequately or on a professional level except by someone with a good working knowledge of hydraulics, thermodynamics, mechanics of materials, and related subjects. An applicant cannot be considered to have developed to the professional level in this field until she or he understands the effect of varying loads on equipment.

A person who has spent many years in engineering may not qualify as a professional engineer. Experience as a drafter, to be qualifying at the professional level, must include a considerable amount of varied and complex design experience. Experience as an apprentice drafter or as a detail drafter is not qualifying. Detailing and other routine work is not qualifying, in accordance with the various state boards' strictest interpretation.

TITLES ARE NO CRITERIA

Salaries and titles are not necessarily good criteria for determining professional experience. Some drafters doing high-grade routine work, but still subprofessional work, make good salaries. The same may be said for those doing routine engineering work in desk jobs. Some drafters and "engineers" become highly proficient in the use of handbooks and other devices for speeding up production design work without ever attempting (on their own time) to understand the basic principles on which such tables and shortcuts were developed. However, some persons designated as drafters are, in fact, doing professional work. The boards' determination about professional experience is based on the evidence submitted as to the exact nature of the duties performed and responsibilities assumed in each case.

RESPONSIBLE CHARGE

"Responsible charge" may mean "of work" or "of people." Work assigned by a superior but not requiring that person's direct and continued supervision may be classified as in "responsible charge" on the application form. A subordinate or subprofessional, in this particular line of thinking, would require continuous supervision and close checking in all phases of the work. The applicant who can attest to and prove the fact that she or he works under her or his own direction definitely qualifies. See the Appendix for more detailed definition of responsible charge.

SPECIALISTS NOT LICENSED

The boards do not license the specialist; they license the engineer. As we indicated before, the applicant's experience must be broad in scope, and diversification must be present to a great extent. The applicant must distinguish routine from real responsibility. Simple operating experience or running of calculations or computer operating (not programming or setting up the formulas) which can be accomplished by persons with no engineering experience or training are looked upon by the boards as simply routine. Thus, experience must be of sufficient breadth and scope to assure that the applicant has attained reasonably well-rounded professional competence in the basic engineering field, as well as highly specialized technical skill in a very narrow and limited branch of his or her field.

WORK AS A CONSTRUCTION CONTRACTOR

The increasing number of engineering-college graduates entering the field of construction employment has raised another question of qualifying experience. Such experience must be of a character that would qualify an applicant to practice engineering independently.

According to the various state boards, the basic principle underlying the practice of professional engineering is *design*. Most contractors in the field of construction execute work previously designed by engineers; and while contracting work may provide the individual employee with knowledge of construction methods which will make that person a better designer, it does not, per se, increase knowledge of design to a degree that would justify its being considered as qualifying experience.

Exceptions, of course, would include the work of an individual who had had the responsibility and charge of construction planning for a contractor on a project such as Boulder Dam or Grand Coulee Dam. Such work could be adjudged engineering experience of a high order and qualifying in nature.

Technical supervision of construction tends to expand the engineering knowledge and skill of the applicant. This can be considered qualifying experience. All boards use their own discretion in evaluating such experience.

Some people feel that the tendency of the boards not to credit contracting work is detrimental to the construction industry. For example, an engineer-in-training now normally takes a written examination upon graduation from college to determine design knowledge and fundamental knowledge. The EIT may take a job with a contractor, and will probably work for six or seven years before any contractor would give him or her any supervisory job whatsoever. If experience time starts from the time the person becomes superintendent, he or she may be twelve or fourteen years out of school before it's possible to go back before the board and take the professional part of the examination. This does work a hardship on the license-minded person.

However, there has occurred a remarkable change in the attitude of the construction industry toward engineers. There is expected to be a continued expansion of employment of engineers in that field.

It often happens that the engineer-in-training wants to know whether work with the contractor constitutes professional or subprofessional experience. If the contractor employs professional engineers, that is a good sign, because the young engineer will be working in a professional-thinking environment. Otherwise, with no professional engineers around, the EIT would be taking on responsibility beyond her or his years and experience. And, in addition, the welfare of the public would be in jeopardy.

Graduates in civil engineering who intend to go into structural design cannot be good designers unless they have been on the job and have seen how the various structural members are assembled. Engineers do not get such valuable experience in college; there isn't time for it. Nor can they obtain the type of experience necessary in an office or drafting-room atmosphere without professional guidance.

The Committee on Qualifying Experience recommends to all boards that, as far as possible, "work as a contractor be considered as experience qualifying an applicant for admission to examination for registration *when* such experience in the opinion of the Board has involved responsible supervision of a character that will tend to expand the engineering knowledge and skill of the applicant; in which event the board may, in its discretion, give such credit therefor as it may deem proper."

VIEWPOINT OF THE EMPLOYER

The viewpoint of the employers of professional engineers has been investigated. Their opinions were collected concerning what the qualifying requirements for employment at various levels of responsibility should be and how to evaluate the knowledge and experience of an applicant. Some 300 questionnaires were sent out to a wide variety of manufacturers, consulting engineers, government and state agencies, constructors, and railroads. General questions were asked, and the condensed answers were found to be of interest. (See Table 7.1.)

Briefly summarized, here are some of the questions and answers:

QUESTION: In selecting personnel, what is your normal procedure; how do you determine and rate qualifications?

ANSWER: The general procedure is based on the interview and record of qualified experienced, supported by references. Governmental agencies, as a rule, use civil service methods, setting up an eligible roster, with appointments on a probational basis. Many firms obtain personnel by training young graduates, so as to have a reserve pool of employees for filling positions by merit promotion.

QUESTION: Discuss the college graduate versus the nongraduate as an applicant for a position.

ANSWER: College graduates are usually preferred, because of their training and potential capacity; however, graduates occasionally expect more pay as beginners than they can produce in value of performance and may have difficulty getting their feet on the ground. Technical education, if not combined with good judgment, is a handicap. College graduates have a foundation upon which to build and are trained to realize their limitations. The nongraduate has a professional peak.

QUESTION: Do you recognize the registration laws of your state? Do you show any preference in employment to those who are registered? Would you venture a suggestion as to how registration could be more beneficial to the engineering profession and to you, the employer?

TABLE 7.1

Employer	Number of firms answering	Employees represented
U.S. government agencies	3	5,000
Constructors	5	3,912
Consulting engineers	50	4,288
Railroads	5	785
Teachers (colleges)	6	506
Manufacturers	24	20,298
Public utilities	1	100
City governments	28	2,453
Highway departments	25	17,048
Total	147	54,390

SOURCE: NCEE (1956).

ANSWER: Registration laws seem to be recognized as a requirement for persons in key positions; and registered engineers are given preference, although registration is not an indication of ability and qualifications. The purpose of the registration laws is to protect the public, but some states seem to restrict practice to their own boundaries, which is not beneficial to the public. Basis should be established for more reciprocal agreements and registration in all states without unreasonable inconvenience, embarrassment, and expense. Having a license is usually a point in the applicant's favor; it is a mark of distinction and accomplishment. But the present laws need more policing, for real benefit to the entire engineering profession.

QUESTION: Do you think the methods used as a yardstick of minimum qualifications for registration in your state are adequate and fair?

ANSWER: As a rule, the answer is yes, but certain arbitrary requirements of examinations seem to be unreasonable and unnecessary. They are satisfactory for the young graduate, but almost impossible for the older engineer. Written examinations alone are not adequate to determine qualifications.

The thinking expressed in the above survey persists to this day. Employers of salaried professional engineers in industry still feel the present state registration laws need more policing, for real benefit to the entire engineering profession; the laws should have more "clout."

A survey of companies conducted by the NCEE Committee on Communications and Publications (Ad Hoc) as given in the *NCEE 1985 Proceedings* showed that many companies operate under the "industrial exemption" provisions in state statutes and do not require their engineers to attain licensure. Most personnel managers (88 percent) feel that licensure should not be mandatory for all engineers in all fields. Firms that provide engineering consulting service to the public do give hiring preference to engineers who have passed the FE examination and they encourage them to pursue PE licensure after completing the required experience.

In the academic community, the survey showed that many academicians see no motivating factor for attaining licensure, since licensure does not contribute to advancement of their career. Only one-third of the persons who take the FE exam continue to pursue licensure by taking the PE exam after they have completed four or more years of experience.

RELATIONSHIP OF EXPERIENCE TO LICENSURE

Licensure is certification of minimum requirements in advance of practice, which develops with time and experience. The certification process is realized by examination following statutory experience requirements. Examination of experience has been found from the very beginning of licensure to be the simplest and most practical way. And until better ways are developed, consideration of practical experience followed by written examination will remain.

Engineering is a glamorous profession, and naturally the esteem and prestige many outstanding engineers bring to the profession, as well as the prestige inherent in the profession, attract many—some of whom are qualified and some of whom are not—who want to use the title "professional engineer." These persons may be good citizens and neighbors, but many are not in possession of adequate technical knowledge and experience in engineering works.

The engineering profession is facing a particular hazard at this time with regard to the maintenance of high professional standards. The consequent recruiting of persons trained at a lower level to do specific engineering or limited technical jobs under the title of "engineer" will cause many with good knowledge in some limited engineering field to think they are professional engineers and entitled to registration. Actually only a few who have such limited experience will have both the ability and fortitude to become licensed by the difficult route of self-study and written examination.

Part Science and Part Art

Engineering practice is part science and part art. The science is acquired by the completion of a registered program. The art is obtained through practical experience. Rarely is a recent graduate prepared for unrestricted and unsupervised practice; so the new engineer is expected to serve an internship under the preceptorship of older and qualified engineers. This is for the purpose of developing and maturing engineering knowledge and judgment. This period consists of practical engineering experience of a grade and character satisfactory to the board of registration.

The practical experience requirement is generally 8 years,* during which the individual adds some engineering thinking to the job—such as making minor decisions on design features, specifications, methods of procedure, and the like. The new engineer does not have to be responsible for the final design but she or he does have to contribute ideas based on experience and engineering fundamentals. This requirement is tested in two ways: by written examination and by personal testimony from those who are familiar with the person's engineering work.

Other qualifications usually checked by examination are English

---

* Four years if after an ABET-approved curriculum graduation or board-approved equivalent.

reading ability, report writing, and ethics. Character references are required. These are consulted to determine the suitability of the candidate for professional recognition. All these factors taken together form a comprehensive and quite dependable measurement of the competence of the individual to practice professional engineering. It is possible to attain this stature only through years of practice of the profession of engineering in a subordinate capacity with constant devotion to personal development through continued education as well as through experience.

At the end of the internship, the candidate may be ready for practice. Certainly, the person still has a great deal to learn, but, at this point, the individual's sense of responsibility and judgment should be sufficiently developed. However, you can't always be sure. The board tries to determine fitness to practice by giving a written examination. The board is concerned with *how well* the individual can apply what he or she knows rather than with what or how much the person knows. Therefore, the written examination is an open-book one, designed by means of well-selected samples to test the candidate's fitness to practice. In my opinion, as the result of working with engineers from industry, government, and consulting practice, there is no better way to test fitness than to consider a person's education, experience, and performance in a written examination requiring the solution of problems related to those the person is being licensed to solve and which affect public safety.

Certification of Book Learning

The undergraduate degree is certification of book learning only. The professional engineering license also attests to practical experience and, in a sense, is the important goal for which the engineer strives. Registration is the fulfillment of an engineering education.

Certainly, the FE examination procedure should stress the degree to which applicants are able to pull together the principles and background information of their engineering education in solving real-life problems in engineering practice. The significance and importance of "reality" in engineering education need constant reiteration. The examination requirement is one means of accomplishing this objective. Compromising our ideals for the sake of gaining membership in the profession is in direct violation of the fundamental purpose of our very existence.

Graduation from an ABET-accredited engineering program does not necessarily ensure that students have retained or assembled the fundamentals presented to them in such a manner as to enable them to apply those fundamentals adequately and competently.

Many have observed "cafeteria" students: They select what they want, pay for what they get, and feel under no obligation to anyone. Such persons not only lack a professional attitude and spirit but do not even comprehend it. Some of these students take courses as a runner takes hurdles—each one out of their minds as soon as it is past. Often, such students accumulate enough credits to graduate, even in an ABET-accredited program, without having ever perceived the unifying structure, the interrelations which underlie not only our physical universe but our profession as well. With all their education, most of them find that the Fundamentals examination is a brand-new experience, though many who fail to pass the first time will pass the second time.

Cafeteria students are unlikely to produce creative, imaginative synthesis or design. To label them as intern engineers without some comprehensive test would, in my opinion, not only fail to adhere to the standard but would violate the spirit and intent of the licensing and registration statutes.

Registration, condensed into one brief sentence, is a screen which should pass freely those who prove minimally competent and exclude those who prove incompetent. No profession can gain respect unless the minimum standard is sufficiently high and registration provides the only path to professional status; however, excessively high standards would be undesirable and illegal, lest engineering become an exclusive club for the "eminent." Then engineering registration would become "class" legislation, and any protective law would become unconstitutional. No registration should carry past the point where it permits the competent to work and prevents the incompetent from deceiving. Many state laws do not require graduation from an accredited program in an engineering school. At present the minimum requirement is high school graduation, but there are forces which are advocating graduation from accredited engineering programs as a minimum. I do not agree with this approach because, if that were rigidly adhered to, we would stand to lose the services of many outstanding engineers who today are making engineering history despite the handicap of an inadequate formal education.

The point here is that engineering requires art and intuition as much as science and formal procedures. The formally educated engineer without the art and intuition is as weak an engineer as the practical mechanic who has inadequate theoretical training. The professional engineering license is the meeting ground for these two extremes of background. The talented, practical, experienced, empirical engineer without a degree, who is willing to study widely enough to pass a broad though still practical professional engineering examination, must be welcomed into the professional ranks as readily as a highly theoretical person, without those with more formal backgrounds feeling that the professional engineering license has been degraded.

Educational institutions judge students' competence almost entirely on their ability to pass tests. Only a small fraction of a student's grade is dependent on willingness to work, on personal ethics, or on leadership qualities. Boards of registration, on the other hand, strive to evaluate the character of experience and the references submitted by the applicant's contemporaries. The boards also require a written examination to test real-life situations such as cost, engineering time, degree of justified precision, etc.

In preparing the Principles and Practice questions, NCEE devises questions that will require, in addition to a solid academic background, those qualities of judgment that can only be gained by good experience. In order to introduce elements of judgment into their examinations, boards use subjective questions devised with the help of NCEE and the engineering community. True or false or objective multiple-choice questions are not used on the PE exam (as they are on the FE exam). Of course, subjective questions are more difficult to grade, but they give the best results and are fair.

ELIMINATION OF THE EXAMINATION WOULD BE DISASTROUS

Certainly, there are those among us who would like to see the PE examination and even examinations in college eliminated. Students and nonregistered persons would like this. But such actions would prove disastrous. Admittedly, tests are not perfect, but they are the best tool available to judge a person's proficiency in a particular subject in a short period of time. A comprehensive examination helps to eliminate those persons who study to pass a course and then erase the subject from their minds. Colleges rely heavily on the final examination to attempt to eliminate the student who does not have a comprehensive understanding of the entire course subject.

The design of structures, processes, and products of all kinds—the output of the engineering mind—is a grave responsibility. A competent practitioner needs more than technical skills to measure up to that responsibility. Engineering is now so complex that a fundamental engineering education is essential to provide the best engineering services to either employer or client.

For the self-made person, the mere passage of time in the engineering office or in the field or in design will not qualify for licensure. That person must have performed tasks reflecting the use and application of engineering principles in a progressive manner and be able to attest to that fact. The use of handbooks to arrive at design criteria without knowledge of their synthesis or theoretical composition is not qualifying.

Such an individual can easily be separated from the college-trained, competent individual by a simple series of tests.

Some unlicensed persons who think they should be registered tell us how much better engineers they are than some of their registered associates. These people can always prove by examination their "right" to be licensed.

ISSUES

Experience before examination. The examination can show how well the engineer has put time on the job to good use in developing engineering skills. Merely performing a job is not the whole answer. The knowledge and skill gained by performing that work is the experience the boards look for and test. Incidentally, this is one way of upgrading the engineering profession.

Relationship of experience to educational requirements. The formally educated engineer without art and intuition is as weak an engineer as the practical mechanic who has inadequate theoretical training. The PE license is the meeting ground for these two extremes of background.

Evaluation of experience after examination. Since the board tests for knowledge and skill gained by job performance of high grade and character, such an examination would be senseless. The person who already has a license but who lacks job knowledge and skill would present potential dangers to the practice of the engineering profession. Such a situation leaves much to be desired since anyone can front for the applicant in the documentation of experience.

Full examination. This provides the means of controlling the developing years of the graduate. It also provides a goal for the nongraduate who can look upon the license as equivalent to a degree in engineering, since she or he sat for and passed the same examination as the engineering graduate. However, the engineer who can attest to qualifications of eminent practice should be considered for waiver of a part or parts of the written examination.

What does examination have to do with experience? The examination is a test of the character of the candidate's experience. Has the experience been broad? Has it been of high quality and character? Has it been diversified? Although there is a close relationship between job performance and experience, a distinction should be made between the two. Job performance relates to the quantity and quality of the work performed, whereas experience is the knowledge or skill gained by performing that work. The quality of experience that one obtains is directly proportional to the quality of job performance. The board evaluates the quality and quantity of the work performed as documented in the application; it tests for the knowledge and skill gained from work performed.

8

CRITERIA FOR EVALUATING EXPERIENCE

There seems to be a wide divergence of opinion about what education is required for the successful performance of duties in professional engineering positions, and the type, quantity, and quality of experience that should be considered as qualifying. The line of sharp demarcation between subprofessional and professional experience, to be recognized as qualifying experience, may vary considerably under different conditions of education. However, it is axiomatic that all persons filling professional engineering positions should be able to perform, completely and successfully, all of the duties of the position and maintain the standards of performance required for the position. It is extremely difficult to prescribe precise criteria whereby the qualifications of applicants can be measured and determination made as to their ability to perform professional engineering work. The information that the applicant gives should be primarily related to the particular branch of engineering in which the applicant seeks registration. Thus, a person presenting background for registration as a chemical engineer would be expected to show a greater and more detailed knowledge of chemistry than an applicant for registration as a civil engineer. By the same token, that person would not be expected to have a great deal of technical knowledge in civil engineering.

EVALUATING EXPERIENCE RECORDS

Some difficulty was reported in 1950 among boards. One board would assume that its practice was perfect and should be the procedure for all other boards. Others would "bristle" over the prospect of infringement of their states' rights. Still others objected to any process restrictions, holding that each board should develop its own procedure regardless of that of any other.

Accordingly, in an attempt to develop some degree of uniformity, the NCEE Committee on Uniform Laws and Procedures suggested a set of standards for evaluating experience records submitted by applicants for licensure, such that, if practiced by all boards, part of the registration procedure would present, in effect, a united front. It made clear that the adoption of these standards would not bind any board to follow them.

The following points of evaluation should give insight into usual board requirements and should prove of great benefit to applicants in writing up their experience records.

A. No experience gained prior to the age of eighteen to twenty should be counted.

B. The subprofessional experience of a technical graduate normally should be rated higher than that of a nongraduate for the same length of time.

C. To be admitted to a written examination as a professional engineer an applicant should be qualified as follows:

1. The applicant should have his or her EIT and, in addition, should have had at least four years of experience as follows:
    a. Three years of experience in professional engineering involving design
    b. One year of experience showing the applicant to be qualified to handle a professional project via responsible charge*
2. If the applicant is a graduate of an accredited engineering program but does not hold a certificate as Engineer-in-Training, he or she should show the same length and quality of experience as the holder of such a certificate.

In the evaluation of experience records, it should be borne in mind that computer programming is considered professional experience.

ABET has proposed a definition of *design* that may be used as a further guide:

> An engineer is characterized by his ability to apply creative scientific principles to design or develop structures, machines, apparatus or

* See responsible charge definition in the Appendix.

> manufacturing processes, or works utilizing them singly or in combination; or to construct or operate the same with full cognizance of their design, and of the limitations of behavior imposed by such design; or to forecast their behavior under specific operating conditions; all as respects a specific function, economics of operation and safety to life or property.
>
> The function of design is not limited to machines and structures. Process design may be defined briefly as the determination of the best process to accomplish a given end from the standpoint of economy, safety, raw materials and available equipment.

It must be clearly understood that this definition was to be broadcast to the profession for discussion. There will be some difference of opinion among boards of examiners as to just how to interpret this definition of design, but it provides a basis for discussion.

DEFINITIONS OF EXPERIENCE

Persons engaged in engineering work may have their experience classified as *professional* and/or *subprofessional* by their board of examiners. In an all-out effort to make this classification procedure more uniform and to give the boards something to "hang their hats on," NCSBEE* instructed the qualifications for registration committee to compile a set of definitions of the types of experience that should be classified as professional and subprofessional. It must be emphasized that these definitions do not relieve an individual board from using its own discretion in evaluating the experience record of an applicant; however, it does indicate the opinions on the subject of a majority of engineers replying to questionnaires. Also, the number of special fields in which this experience is gained need not be limited. The definition simply indicates that experience in any one of the listed subfields can be considered as qualifying experience. Remember, the listing is far from complete.

Civil Engineering

Professional experience

Design and/or supervision of construction of: highways; streets; subways; tunnels; drainage; drainage structures; sewerage; sewage dis-

* Now NCEE.

posal; structures; railroad spurs and turnouts; railroads; rivers, harbors, and waterfront; waterworks; water supply; water power; airports and/or airways; bridges; dams; irrigation; irrigation structures; water purification; incinerators; storage elevators and structures; flood control.

Development of: structures for regions; ports; wharves and docks; irrigation works; flood control; drainage.

Investigations; appraisals and evaluations; cost analysis; consultations; testing; research and economic studies; reports.

Teaching at college level, full-time, in an accredited college of engineering.

Editing and writing engineering subjects.

Surveys: topographic; hydrographic; earthwork; triangulation; railroad; irrigation; drainage; water power.

Independent responsibility and supervision of the work of others.

Subprofessional experience

Fieldwork: instrument operator; chainman; rodman; chief of party; recorder; surveyor; leveler; staking out foundations; staking out dredging ranges; railroad lines and turnouts; sounding; instrumental observations.

Drafting: plotting field notes; map drafting and tracing; tracing designs; structural detailing; plotting soundings.

Construction: inspection; timekeeping; supervision without authority to change designs.

Computations: routine, under direction; taking off quantities; monthly estimates.

Cost data: routine assembling and computing; estimating under direction.

Testing: routine testing of materials of construction; water.

Teaching: as an associate without full responsibility in an engineering college.

Mechanical Engineering

Professional experience

Design and/or supervision of construction of: machines; machinery and mill layouts; heating, ventilating, and air conditioning; power plants; power-plant equipment; industrial layouts; refrigeration; tools and processes; designs involving application of existing equipment and principles; internal-combustion engines; rocket engines; CAD/CAM; robotics:

Development of industrial plants and processes; consultation.

Appraisals; investigations; research and economic studies; cost analyses; reports.

Operation: major mechanical installations; manufacturing plants; power-plant testing; application of instruments and control to manufacturing processes.

Editing and writing on mechanical engineering subjects.

Teaching, full-time, at college level, in accredited schools.

Subprofessional experience

Construction and installation: machinery; heating and ventilating; air conditioning; mechanical structures.

Operation: heating, ventilating, and air-conditioning equipment; power plants; mechanical manufacturing plant; stationary machinery (small operations); foundry and machine shops.

Drafting: detailing; shop drawings; checking shop drawings, tracing; layout work. Design of tools, jigs, and fixtures.

Recording; routine computing under competent direction; laboratory assistant; routine testing under competent direction; inspection of materials, etc.

Maintenance; repairs; welding.

Teaching as an assistant without full responsibility, in an accredited college.

Electrical Engineering

Professional experience

Design, specifications and/or supervision of construction: power plants; electronic and transmission equipment; application of electrical equipment to industry; electric service and supply; communication systems; control systems and servomechanisms; manufacturing plants; operating procedures; distribution networks; transportation; radio circuits; major electrical installations.

Development and production: new techniques and devices; electrical and electronic devices.

Supervision and/or management: power plants; manufacturing plants.

Research and investigations in electrical fields; water, steam, and electric power; electronics; new techniques and devices; high-voltage cables.

Consultation; appraisals; valuations; estimates; cost analysis; reports; determination of rates; economics and cost figures; calculations of system and device performance.

Sales and application of equipment to industry; service engineering; testing, when not of a routine character.

Editing and writing on electrical engineering subjects.

Teaching, full-time, at college level, in an accredited school.

Subprofessional experience

Construction of electrical installations, under competent direction; electronic and radio equipment; wiring; lineman; switchboard wiring; electrician; meterman; armature winding.

Inspection: routine, under direction.

Installation: heating, ventilating, and air conditioning; radio equipment.

Drafting: elementary layout; circuit drafting; detailing.

Testing: routine; performance of electrical machinery.

Operation: radio; power stations; substations; switchboards.

Repairs: motors; generators; electrical devices; electronic and radio devices.

Maintenance: electronic and radio devices; radio and television technician.

Apprentice: work as a junior in fields of illumination, communication, transportation.

Sales: routine; almost all customer service; inventory.

Teaching: manual skills; instructor in accredited school.

Chemical Engineering

Professional experience

Design, specifications and/or supervision of construction: equipment for chemical enterprises; processing plants; plant layouts; equipment;

economic balances; production planning; pilot plants; heat-transmission apparatus.

Development: processes; pilot plants; refrigeration systems for food processing.

Research: laboratory; pilot plants; director; markets; unit operations; kinetics; processes; calculation and correlation of physical properties.

Responsible charge of broader fields of chemical engineering; executive.

Consultation; appraisals; evaluations: reports; economics; chemical patents laws.

Operation of pilot plants; product testing; technical service or sales.

Teaching, full-time, at college level; editing and writing.

Subprofessional experience

Construction: process equipment; pilot plants; piping systems.

Operation: shift operator; special chemicals manufacture (solutions); pilot plants; trouble shooter; glass blower.

Drafting: flow-sheet layout.

Instrument making and servicing: routine analyses; routine sampling and tests; latheman.

Analyst: routine, under direction. Laboratory assistant (commercial, college); computations, under direction; data taking.

Sales: routine, of standard equipment.

Aeronautical Engineering

Professional experience

Design, specifications and/or supervision of construction: structures: power plants; wind tunnels; aircraft; missiles; helicopters; propellers; controls; layout; loftsman.

Research: wind tunnels; flight; structures; stress analysis; aerodynamics; instrumentation.

Subprofessional experience

Flight instructor; wind-tunnel operator; wind-tunnel model building; test pilot; engineer testing; drafting; lofting; aircraft maintenance; aircraft inspection.

Airport operation: drafting (under direction); inspector; A & E mechanic. Some phases of testing of a routine character with no research or analysis required.

Agricultural Engineering

Professional experience

Design, specifications and/or supervision: agricultural machinery: research; drainage; irrigation; surveys; terracing; farm electricity.

Teaching, full-time, at college level, in accredited school.

Subprofessional experience

Drafting; wiring; equipment installation or servicing; routine inspecting; dealer or consumer training; surveying under direction; farm-machinery operator.

Mining and Metallurgical Engineering

Professional experience

Design, specifications and/or supervision of construction: plants; shaft and bottom layout; advance layout of mine workings; ventilation; haulage; hoisting systems; headframes; washery or concentration mill; designs involving application of existing equipment; production planning for manufacture of metals or metal objects; mining methods; design and use of mining machinery; testing procedures; metallurgical works; preparation-plant designer and controller.

Consultations; mine examinations; engineer in charge; superintendent; positions requiring judgment where decisions are final; ore estimation; mine safety.

Geological studies of an exploratory nature; control of mine surveys; Federal Bureau of Mines inspection.

Mine administration, including responsibilities as shift boss, captain, superintendent, manager; mine or plant engineer; plant-maintenance engineer; mine-ventilation and mechanization engineer; control of mine surveys; control of sampling; guidance or mine development; responsibility for the efficient and economic production of metals, alloys, or metal objects; responsibility for the selection of materials, processes, and treatments.

Research; head analyst; chief metallurgist; valuation; beneficiation studies; time studies; development of alloys; mineral land explorations and appraisals; supervision of processing.

Inspection of mines.

Teacher, full-time, at college level, in an accredited school.

Sales engineer, above the district level; purchasing agent.

Subprofessional experience

Assaying; sampling; foreman; mill operator; surveyor; plotting notes; running levels; drafting; tracing; inspecting construction; mapping; detailer; topographer.

Surface and underground construction; installation of machinery; instrument repair and service.

Inspector: safety; ventilation; time study; routine testing; stope geologist.

Shift boss in mine, mill, or metal plant; mucker; miner; timberman's helper; jigger boss; trainee; metallographer; sampler; heat treatment; mechanical testing; laboratory assistant; furnace operator; routine washery or mill chemist.

Cost accounting; routine geological recordings; estimator.

Sales; shipping and transportation.

Teaching, part-time, no responsibility.

Petroleum Engineering

Professional experience

Design, specifications and/or supervision of construction: drilling equipment; pipelines; refinery plants; gathering systems; handling and storage.

Choice of location of wells; drilling engineer; mud engineer; development engineer; subsurface engineer; proration engineer; corrosion engineer; oil and gas reservoir performance prediction; research and development of petroleum products; cost engineer; exploitation engineer; field engineer; operations engineer; subsurface geology; reservoir measurements; exploitations; tool pusher.

Research; exploitation; valuations; selection and installation of production equipment; primary and secondary recovery of oil and gas; logging of oil and gas wells; coring and core analysis; measurement of gas; studies of phase behavior; reserve calculations; consulting engineer.

Teaching at college level in accredited schools; geologist; geophysicist.

Subprofessional experience

Operation: oil and/or gas wells; pipelines; refiners; field roustabout; roughneck on a drilling rig; scraping paraffin from wells; pumper; driller; gauger; tool pusher.

Equipment installation.

Switching and gauging; obtaining bottom-hole pressures; mechanical tests; production tests; routine tests.

Map drafting; farm boss; assistant to engineering classifications given under professional experience.

Teaching, part-time, under direction; routine computer; student engineer or apprentice.

Traffic Engineering

Professional experience

Design of traffic islands; design of new and augmented control systems; design of relief systems; design of highway lighting; traffic-engineering planning and roadway design; design of mechanical traffic control.

Planning and analyzing surveys and traffic counts; interpretation of field data; computing earning power of proposed routes; establishing traffic-count stations; transportation studies (major or complete); traffic-education programs.

Traffic-engineering organization and administration. Teaching at college level.

Subprofessional experience

Traffic counts; compilation of data; drafting; timing of automatic signals; plotting traffic-flow charts; plotting traffic-dispersion charts; origin and destination data; weighing stations; accident study and tabulation; sign and signal checking.

Naval Architecture and Marine Engineering

Professional experience

Design, specifications and/or supervision of construction: hulls; marine machinery; propellers; piping; marine electrical equipment; boilers and heat transference; structures; heating and ventilating; machine-shop and marine-ways practice; cargo handling; docks.

Calculations for stabilization; testing of hull forms in naval tank or model basin. Teaching, at college level.

Subprofessional experience

Mechanical drafting; strength of materials; preliminary electrical engineering; estimating and costs.

Industrial Engineering

Professional experience

Administration: manufacturing plant; wage and salary; any major supervisory position; plant manager; production manager; maintenance engineer; plant engineer; tool engineering; operations supervision.

Design: factory layout; tool and fixture; factory planning; motion- and time-study program; setting final standards and methods; rate plans and related work; development of production plan; development of quality-control systems; quality-control-systems analysis; methods-time measurement; work simplification; method study; machine-loading analysis; job analysis; cost analysis; production-control analysis; organizational analysis; process analysis; economic analysis; operations analysis; labor-needs analysis; safety engineer.

Research; teaching, full-time, at college level in accredited school.

Subprofessional experience

Production-control clerk; expediter clerk; materials-control clerk; inventory-control clerk; control-chart clerk; cost clerk; labor time-card clerk; material-cost clerk; time-study clerk; inspector; dispatching and following up; stockroom clerk; "stop-watch-clipboard" data taking; millwright clerk; scheduling clerk; rate setting; job rating.

Process and flow-chart construction; equipment installation; job fitter; layout; maintenance inspector; machine setup; routine shop work. Superintendent; drafting; detailer; die designer; routine motion study; machine scheduling and machine records; routine production scheduling.

Geodetic Engineering

Professional experience

Selecting technical methods; outlining mathematical procedures; adjustment of difficult schemes of second- and third-order triangulation; computer and adjuster of involved triangulation schemes; reviews computations of lower-grade computers.

Checks astronomic azimuth computations; reviews tabulated results of triangulation and traverse computations; solves miscellaneous problems pertaining to rectangular coordinates, map projectors, and geodetic azimuth.

Layout of base lines; layout of triangulation networks; calculation of geographical positions and map projections.

Teaching, full-time, at college level, in an accredited college.

Subprofessional experience

Checking notebooks for errors; adjustment of secondary traverse and level lines using mathematical tables and calculating machines; routine calculations.

Chief of survey party; rodman; signalman; observer; drafting; tracing.

Sanitary Engineering

Professional experience

Design, specifications, and/or supervision of construction: water-treatment plants; sewage-disposal plants; waterworks; sewers; incinerators; distribution systems; sewage- and industrial-waste-treatment plants.

Consultations; reports for proposed works; original research for processes and equipment; review of plans of works for state and federal authorities; planning and analysis of surveys and pollution data; advising on water-purification plants; counseling municipal officials concerning needed sanitary facilities; planning and supervising environmental sanitation work in sanitation of milk, food, etc. Investigations; studies; reports and evaluation of proposed or existing water or sewerage systems; industrial-hygiene surveys and reports; research in sanitary engineering or public health.

Abatement procedures on stream pollution; supervision of pollution surveys or of works for large cities; control of rodents and insects.

Teaching, full-time, at college level, in an accredited school.

Subprofessional experience

Routine laboratory tests; installation of machinery; gathering records for reports; stream gauging; surveying; plant operation and maintenance; record keeping; sanitarian's inspection; mapping; inspection; routine chemical and bacteriological tests; pest control; collection of data; take-off; sampling; establishing of lines and grades; inspection of shellfish plants; routine inspection of eating and drinking establishments; routine inspection of vessel sanitation; routine watering-point inspections; orientation periods in public-health agencies; inspection of construction.

Maintenance and operation of filter and sewage plants; routine laboratory examinations.

Structural Engineering

Professional experience

Design, specifications, and/or construction on structures; hydraulic design; stress analysis; soils analysis; mechanical analysis.

Research on structures; structures testing; layout studies and comparisons; new design procedures and methods of analysis; reports; consultations; investigation evaluations.

Project supervision; field execution of projects; senior structural draftsperson; senior stress analyst; senior engineer on materials testing; senior concrete laboratory technician; resident (field) engineer on bridges, dams, buildings; research engineer.

Teaching, full-time, at college level in an accredited school.

Subprofessional experience

Structural drafting, detailing; drafting layout and details; junior stress analyst; computer; checking detail drawings; checking design computations; collecting of data for reports; inspector; junior laboratory engineer; research assistant; junior concrete technician.

Electronics Engineering

Professional experience

Design, specifications and/or supervision of construction: design engineer; design and development of electron tubes and apparatus; radio transmitters and systems; all types of receivers; antennas; facsimile; radar; broadcasting stations; components and systems; design of important systems; equipment planning for major manufacturing operations; extension into new fields; nuclear work; supervision of major installations; telephone systems; television equipment; communication relaying systems; public-address systems; medical equipment; filters; servomechanisms; motor and industrial controlling devices; railroad signaling systems.

Major responsibility for research; analysis of fields of application for control, heating, etc.; executive in production or application of electronic equipment.

Teaching, full-time, at college level, in accredited school; major technical papers and tests.

Subprofessional experience

Same as electrical engineering.

Wiring; assembling; routine testing; operation of equipment; drafting; glass blowing; repair and maintenance; assembly and construction of electronic devices; repairperson; installer; factory foreman; installation, operation, and testing as technician with electron tubes and apparatus; radio servicing; trouble shooting new circuits; telephone-line work.

Assistant in research, development, design, as of conversion, control, measurement; supervision of installation, testing, inspection of systems; "contact" for manufacturers; instructor in engineering school.

Radio Engineering

Professional experience

Design, specifications and/or supervision of construction: transmitters; receivers; measuring equipment; communications systems; major installations; telephone systems; television equipment; communication relaying systems; antennas; filters; complete radio stations; radar; railroad signaling systems.

Subprofessional experience

Same as electrical engineering and/or electronics.

Radio and public-address (P.A.) system operation; supervision of equipment installation; performance tests; inspection; radio-frequency measurements as field-strength surveys; transmitter-location test; field-contact engineer of equipment manufacturers; production engineer for smaller operations; radio station; attendant; station monitor; radio technician; television-camera operator; radio repair; servicing; wiring.

Drafting; assistant in research and development; instructor in accredited school.

Photogrammetric Engineering

Professional experience

Solution of stereophotogrammetric triangulation; stereophotogrammetric mapping; stereophotogrammetric adjustment to aerial triangulation; planning flight map; planning control (vertical, horizontal); specifications.

Responsible charge of subprofessional groups; compiling topographic detail by use of stereophotogrammetric plotting instruments; interpretation of photos for engineering works; design of new equipment.

Subprofessional experience

Transferring existing map information to base sheets by tracing or use of projection machines; plotting geodetic positions; drafting; constructing grids or projections on base sheets; computing projections as assigned for construction; compiling complex planimetry by use of stereophotogrammetric instruments; flight planning; photo indexing; member of survey party.

Architectural Engineering

Professional experience

Design, specifications and/or supervision of construction: structural; architectural; heating and ventilation; mechanical equipment; plumbing and water supply; buildings; repairs and alterations to existing buildings; estimating; appraisals; drawing and letting of contracts; study and selection of proper structural and mechanical system for buildings; aesthetic design.

Consultation; investigations; teaching, full-time, at college level in accredited school.

Subprofessional experience

Labor or checking on construction projects; drafting; routine calculations; shop drawing; shop work; assistant clerk of works; timekeeper; surveyor's assistant; records bulletins; daily reports; progress reports; scheduling; estimating take-offs; expediter; detailing; bar lists; checking shop drawings; inspecting construction; coordinating deliveries; measurements of extras and omissions; job superintendent; checking of specifications.

Ceramic Engineering

Professional experience

Design, specifications and/or supervision of construction: plant and equipment; product development; process development; engineering design; driers; kilns; operational equipment.

Research; consultation; development and analysis of tests on raw materials and finished products; recommendations for products use; fundamental and applied research; raw-materials surveys.

Supervision of plant operation; supervision of mining operations; supervision of construction; industrial supervision and management; laboratory director; statistical quality control.

Teaching, full-time, at college level; sales; writing; preparation of patents.

Subprofessional experience

Operation; testing and control of raw materials; machinery and furnace maintenance; chemical analyses; kiln operator or foreman; sampling and inspection; maintenance of plants and mines; installation of equipment; laboratory assistant; supervision of manufacturing at foreman level; manual work in research laboratory; process and product control; prescribed and routine operations; estimating costs; draftsperson; time study; technical librarian.

Geological Engineering

Professional experience

Consulting work on railroad, highway, and dam construction; building foundations.

Teaching, at college level, full-time.

Subprofessional experience

Mapping; petrography; teaching, as assistant.

Manufacturing Engineering

Professional experience

Planning and selection of the methods of manufacture; development of equipment for manufacturing; research and development to improve the efficiency of established manufacturing techniques; development of new manufacturing techniques; research and development.

Facilities planning, including process, plant, and equipment layout.

Tool and equipment specifications, selection, and development.

Value analysis and cost control; feasibility studies; review of product plans and specifications; research into the phenomena of fabricating techniques; research and development of new production methods, tools and equipment; coordination and control of production within departments; maintenance of production control to assure compliance with scheduling; economic studies.

Subprofessional experience

Operation; testing; machinery and equipment maintenance; installation of equipment; estimating costs; draftsperson; time and motion study; routine operations.

Nuclear Engineering

Professional experience

Design of the specialized equipment and systems of facilities utilizing radioactive materials; analysis and the evaluation of the safety and reliability of the plant or process utilizing radioactive materials to ensure safe operation; design of those aspects of systems for handling, fabricating, transporting, and disposal of fissionable material that involve nuclear criticality and radiation protection; research and development; studies of the nuclear fuel cycle; engineering supervision of nuclear-related tests and operation of facilities; studies aimed at defining the optimum use of nuclear material, and the disposal of waste products, so as to provide maximum protection to the public at minimum cost; provide consultation for engineering groups in regard to any of the activities; management of design groups responsible for the nuclear engineering, safety, and evaluation of facilities utilizing radioactive materials.

Subprofessional experience

Operation and maintenance; routine analyses; equipment installation and routine testing; draftsperson; print coordinator; teaching as an assistant.

Fire Protection Engineering

Professional experience

Research and development; studies for the development of engineering techniques for the prevention and suppression of fire; development, evaluation, and testing of fire detection and extinguishing systems; recommendation as to proper layout and construction of facilities for restriction or spread of fire; evaluation of the fire potential in buildings or structures, groups of buildings, or communities; evaluation of effectiveness of public or private water supply systems, fire departments, and fire communication systems; investigation of the behavior of materials exposed to fire and investigation into the operation of existing fire protection systems; preparation of recommended changes in building and fire codes to keep them current with new methods of construction and the use of new materials. Teacher, full-time, at college level, accredited school.

Subprofessional experience

Routine testing; draftsperson; operation; calculations; cost accounting, surveys, repetitive operations; installation of equipment; estimating under direction. Teaching as an assistant without full responsibility, in an accredited school.

It should be noted here that some state boards holding written examinations in the major professional groupings (civil, chemical, electrical, and mechanical) will classify a person with industrial-engineering experience, for example, as a mechanical engineer and require that that person take the mechanical-engineering examination. This is but one of the obstacles to full reciprocity between states.

Engineering may be defined as the application of mathematics and science to the solution of problems of design, investigation, evaluation, or planning of utilities, structures, machines, processes, circuits, and production. A person becomes competent to do engineering work, first, by studying mathematics and science and their applications and then by acquiring judgment from experience with solutions which have been tested in actual practice.

In all engineering disciplines, the computer and microprocessor have been put to use in many areas of the workplace. Qualifying experience in this area includes the application of the tools of engineering (mathematics, chemistry, physics, and the engineering sciences) to the programming of the computer and the microprocessor.

The mere use of software for a specific engineering task or keypunching is not considered qualifying experience. The following are some samples of qualifying functions in computer usage:

Computer organization and architecture
Computer system design
Computer system applications
Microprocessor system design and application
Data communication systems and networks
Communication networks
Computer-aided manufacturing
Computer-aided processing
Computer-aided design
Computer graphics
Computer programming in FORTRAN, BASIC, Pascal
Application of programming tools and environment
Microprocessor interfacing
Programming of microprocessors

QUALIFYING EXPERIENCE

The definition of qualifying experience given in the previous chapter provides each board of examiners with freedom of action to use its own careful judgment in evaluating any given experience along normal lines. For instance, it should be evident that a detail draftsperson is not making decisions that fulfill the requirements of qualifying experience. Such experience is to be considered only a steppingstone to more responsible work. A teacher who confines her or his work to teaching is not getting qualifying experience. But the nature of the work that comes under the supervision of the teacher, the experience in research and development that may be part of the teaching responsibility, and additional engineering experience in connection with consulting work during the academic year and in the summer periods deserve full consideration and credit in determining the applicant's qualifications.

The definition of qualifying experience does not state in detail how important the engineering decisions must be to be considered professional-level experience. That is the responsibility of the individual state board to decide. Neither does it state the required amount of time to be spent gaining experience. This time is left to the state laws and the judgment of individual boards. *Quality is more important than quantity,* but quality is more difficult to weigh. Thus, most state registration laws specify quantity in terms of time and leave the quality to the judgment of the boards.

Nonqualifying Experience

The following tasks and functions are not considered qualifying experience and may not be included in the experience record when filing for licensure:

- Drafting and detailing; drawing coordinator; routine duties of maintenance, operating, troubleshooting; military service involving routine duties (no research, no development or design).
- Work as a contractor (installation or construction from engineered design drawings and specifications).
- Routine analyses and computations; routine testing of equipment or apparatus; routine testing of materials; repairs and welding; sales without application knowledge; work as an apprentice; inspection; cost accounting associated with or not in engineering works; estimating (tabulation of costs); correspondence courses in engineering not within the community of scholars; operating engineer licensed or not; teaching in a nonaccredited college or school; work during vacation periods between college terms; parts ordering; any employment of a nonengineering nature even though within an engineering firm; parts and material procurement; purchasing follow-up; model making; vendor contact, expediting; bookkeeping; personnel activities; continuing education activities; project reporting; plotting graphs; drawing change orders; mechanical assembly; checking tolerances; procedure writing; requisition writing; routine paperwork; time-keeping; time and motion study.

Teaching as Qualifying Experience

NCEE has drafted a resolution with regard to teaching as qualifying experience.

The resolution *suggests* that the various state boards, when evaluating the experience records of engineering teachers,

> . . . for registration or license as professional engineers, shall require that at least one year of the qualifying engineering experience shall have been in some other area than advanced study, teaching, or research in a college or university, and shall have been under the direction of a registered or licensed professional engineer. . . .

The resolution then goes on to advise

> that administrative officers concerned with engineering instruction in colleges and universities shall require that staff members who teach professional engineering courses be, or be under direct supervision of, registered or licensed professional engineers who have had at least one year of suitable engineering experience after graduation, in some other area than advanced study, teaching, or research in a college or university; and that such experience shall be given equal recognition in respect to advancement in rank and salary year for year, up to at least two years, to the recognition given for graduate study leading toward an advanced degree.

It is the consensus of many that teaching alone without actual engineering practice should not be considered sufficient. It is highly desirable that an engineering teacher should have both experience and advanced degrees, but both cannot be secured at the same time.

Some states do not legally require that teaching be granted credit in registration. The character of teaching and determining the responsibility involved in teaching are still up to the board of examiners. In general, the laws have some such phrase as "experience satisfactory to the board," and most boards will use this as a minimum recommendation. Summer work and experience in industry may count for a proportionate part of the year. It may not be necessary to acquire the full one-year minimum experience outside of academic circles in a single year; this may be made up of experience gained in summer employment and also during leaves of absence. Many faculty members have been told it isn't a question of whether they can afford to get out and get that experience, but a question of whether they can afford not to get it. This need not be a financial hardship. Any way that the teacher is able to meet the minimum of the year of experience will probably be accepted, whether in continuous experience or in several different summers or in part-time industrial experience, so long as this experience is obtained under the direction and supervision of a qualified professional engineer.

Experience Credit

It is within the discretionary powers of the various boards of examiners to allow credit, subject to certain restrictions, for the engineering experience of an applicant.

Most boards *do not allow* experience credit for the following:

1. Teaching of nonengineering subjects in an engineering college or elsewhere.
2. Work projects performed before the age when an applicant would normally complete high school (eighteen to twenty years of age).
3. Assignments in branches of the armed forces which do not involve engineering design, supervision of engineering work, or research.
4. The vacation periods between terms of a college course.
5. Sales work which does not involve the use and application of engineering knowledge.
6. Employment of a nonengineering nature, even though it be in connection with engineering works.
7. The duties of a contractor, superintendent, or supervisor on construction work unless such work involves engineering practice.
8. Correspondence-school courses in engineering.*

Where the duties of an applicant involve performance in both engineering and nonengineering assignments, it is general practice to grant partial experience credit, to the extent warranted by the actual situation. In certifying an experience record, the applicant should carefully designate by description and extent of time involved the actual engineering duties performed, particularly in employment which may not be entirely of an engineering nature.

An applicant may acquire experience in his or her home state or elsewhere. Engineering experience acquired prior to graduation from college may be accepted if it does not fall within the defective categories listed above. A year of graduate study in engineering which has led to a master's degree may be accepted as one year of the four of acceptable experience after graduation. Higher degrees tend to be given less credit in terms of years of acceptable experience.

---

* National Home Study Council, 1601 18th St., N.W. Washington, DC 20009.

Checklist for Evaluating Experience

Here is a useful checklist for evaluating experience. It outlines in detail the high points to bring out in the application.

Experience

1. How much and what kind of technical engineering experience do you have?
2. Does the experience you offer show only the routine use of procedural methods and techniques without regard to their limits or fields of application or the theory involved in their development?
3. Is the experience you are offering as qualifying really not engineering work, but rather that which a highly trained technician or mechanic could be expected to perform successfully?
4. Does the experience offered show an increasing use of the more common scientific principles related to or basic to a particular field of engineering?
5. Does your experience show definitely the knowledge of fundamental engineering in the type of work you have performed, or have you been doing a repetitive job in a routine way?
6. Does your experience record show that you have used, or been given the opportunity to use, scientific approaches to engineering problems?
7. Does the experience you offer as qualifying call for more than the routine observation and recording of technical data and for more than systematized and unvarying operations?
8. Does your record present an experience background that shows more than a passing interest and initiative in your work and does it reflect that you have developed professionally when given the opportunity?
9. Has your experience been narrow and limited in scope, or has it been diverse? (A bit of reflective caution here: diversity of activities may be indicative of inability to cope with difficult problems, or instability, or lack of direction. On the other hand, a person specializing in one line of work may never have had the opportunity to demonstrate knowledge and versatility.)
10. Has your general record of employment been progressive and of increasing responsibilities, and does it indicate progress in the application of engineering principles?

Education (other than engineering)

1. Is this education related to professional engineering? And how much of it directly parallels engineering courses?
2. To what extent have you augmented your knowledge through education at night school, by correspondence, or by home study?

Other qualities

1. Have you shown any interest in other engineering work related to the tasks you have actually performed or are performing on the job?
2. Have you demonstrated an intellectual curiosity toward the basic sciences applicable to your branch of engineering?
3. Does your experience indicate that you have the characteristics of a good engineer—namely, adaptability, initiative, originality, intelligence, capacity to assimilate information?

In a study of what constitutes qualifying experience, NCEE has done much research in the field. Let us look at a typical example of one of their findings:

> An applicant with or without formal education or with limited formal education applies for eligibility in the structural option of a professional engineer examination. What does the board look for in experience?
>
> Problems in structural design cannot be handled adequately or on a professional level except by one with a good working knowledge of the theory of bending moments, shears, stresses, deflections, deformations, mechanics of materials, and of other related subjects. However, an intimate knowledge of chemistry, thermodynamics, electricity, etc., is not absolutely essential, although desirable, to one working solely in the field of structural engineering.
>
> Stresses cannot be computed, however, unless one is first able to determine whether a structure is statically determinate or statically indeterminate. That determination must first be made before the method of approach to the stress analysis can be decided, and structural elements cannot be designed until stresses are determined. Stresses can sometimes be determined graphically, other times through algebraic computation, but in either case the same knowledge of mathematics is involved.
>
> One cannot be considered to have developed to the professional level in the structural engineering field until he understands the effects of varying loads on stresses and how to compute maximum and minimum stresses of a structural element subjected to moving loads, impact loads, and vibratory loads. Only that structural experience which involves

stress analysis, selection of size and type of member, determination of foundation bearing stresses and overturning moments, writing of technical specifications and similar high grade experience should be considered as qualifying. . . .

A person, therefore, who has spent many years in the field of structural drafting may still not qualify as a professional structural engineer. Draftsman experience to qualify at the professional level must include a considerable amount of varied and complex design experience. Experience as an apprentice draftsman or as a detail draftsman should not qualify. Detailing, computing rivets at joints, drafting, and other routine structural work is not qualifying. Neither should construction experience be considered qualifying in the structural option except as mentioned in a previous section on construction. The determination must be based on the evidence submitted as to the exact nature of the duties performed and the responsibilities assumed in each case.

9

WRITING UP YOUR EXPERIENCE RECORD

Before going further in the important matter of writing up an experience record, the applicant should read Chapters 7 and 8 again. This will provide an excellent insight into what the boards are looking for and what the board members need to know about an applicant's experience record in order to give full credit and just weight to the applicant's hard-earned experience.

The application is not the place for modesty, on the one hand; nor is it the place to boast needlessly, because all evidence must be attested to by those the applicant will select as references. These persons will of necessity be most intimately acquainted with the engagements listed on the application form and will have to sign written statements to this effect. Most state boards require the signatures of at least three practicing registered professional engineers personally acquainted with the applicant's professional record. Some states, like New York and Louisiana, require these three signatures to be those of resident engineers.

The experience record must show how much and what kind of engineering experience the applicant has. It must reflect progress toward more important work and greater responsibility on the job. It must call for more than the routine observation and recording of technical data and for more than systematized and unvarying operations. The applicant must show that she or he has had some initiative and more than a passing interest in her or his work and that she or he has developed professionally. Experience should be diverse but complete, not narrow or limited in scope.

It must be remembered that "responsible charge" means independent control and direction, by the use of initiative, skill, and independent judgment, involving complete projects in the applicant's specific field.

APPLICATION MATERIAL REQUIRED

The applicant is responsible for filing all material required by the board of examiners, including the application form itself and a certified education record from the engineering college attended, either as a special form supplied by the board or a certified transcript of college record. All material must be accounted for and indicated on the application form, if space is provided, or in the letter of transmittal.

If the applicant's engineering college is not accredited, the applicant should be prepared to file a *certified transcript* of his or her college record. Applicants who are not graduates of accredited engineering schools but who are graduates of approved high schools or the equivalent or who have partial engineering college credit must be prepared to submit a *certified transcript* of their records covering their education. Extension and evening college graduates are usually required to file transcripts.

Reference Forms

In a manner and on forms approved by the board of examiners, reference forms are endorsed by persons listed on the application form. These forms file certification as to the candidate's moral character, integrity, technical knowledge and competency, experience, reputation as an engineer, and citizenship (or declaration of intention). Such reference forms are normally required to be filed only when full licensure is being sought. For the EIT certificate, such references may not be required.

The applicant should be sure that his or her references understand the questions asked. Below are listed typical questions asked of references. There may be variations, but on the whole, all states follow pretty much the same pattern.

Personal

1. How long have you known the applicant?
2. Have you any interest in the applicant that might influence and color your answers in the applicant's favor, such as friendship or business association?
3. What, if any, are your business relations with the applicant?
4. Have you had any unsatisfactory business dealings or unpleasant experiences with the applicant?

5. Are you acquainted with the applicant's general reputation among your mutual acquaintances as a person of good moral character?
6. Is the applicant a citizen of good moral character?
7. Does the applicant have the characteristics of a good engineer, such as adaptability, initiative, originality?
8. Would the applicant enhance and add stature to the profession by reflecting a high degree of integrity?
9. In your opinion, assuming the applicant knows the difference between right and wrong, would you say that the candidate would actually do what is right and avoid doing wrong?

Professional

1. How long has the applicant been engaged in active engineering work? With which portion of the applicant's work are you acquainted?
2. How long has the applicant been in responsible charge of engineering work?
3. How long has the applicant been in design of engineering work?
4. In your opinion, is the applicant competent to be placed in responsible charge of engineering work?
5. How much and what kind of technical engineering experience does the applicant have?
6. Has this experience been narrow and limited in scope or has it been diverse?
7. Does this experience show only the routine use of techniques without regard to their limits or fields of application?
8. Does the applicant's record of employment show progression into more responsible work?
9. Has the applicant developed professionally by showing more than a passing interest and initiative in his or her work?
10. Has the applicant shown any interest in other engineering work related to the tasks she or he has been performing on the job?

Record of Active Practice

In writing up the record of active practice, the applicant should follow instructions carefully. The applicant should list membership in technical societies or professional organizations. This is an important aspect of professional life and reflects generously in favor of the applicant.

The record should be set down in chronological order, with the first engagement heading the list. All time and types of engagements and number of each should be written down for reference. Do not list experience prior to age eighteen. Dates may be in month and year, i.e., 9/84.

In the experience write-up, applicants should stress engineering design, if it can be attested to. This is sure-fire qualifying experience. Military service, if involving engineering design, development and/or research, is evaluated into years of accredited experience. The candidate should go back and check the evaluation-criteria list in Chapter 8.

Applicants should describe the kind of work done and the degree of responsibility for each engagement. Again remember that "responsible charge" means "of work" and/or "of people."

If applicants must describe a structural project, they must indicate the project by name and location. They must include some descriptive information about size and weight of steel or concrete involved. For equipment design, they must give nature of equipment, such as steam generator (boiler), heat exchanger, pump, refrigeration system, etc.; and they must state rating of boiler (evaporation rate), boiler outlet conditions of temperature, either drum or superheater outlet pressure, and heat-transfer surface and service conditions. For the pump, they must give its capacity, nature and characteristics of liquid pumped, pumping temperature, and total head; for refrigeration, tons of refrigeration.

Any of the following will enhance the applicant's position:

1. Scholastic honors
2. Membership in honor societies
3. Contribution to the technical literature
4. Technical publications
5. Advanced studies and degrees
6. Copyrights
7. Membership in technical committees
8. Patents
9. Important developments and achievements
10. Contributions to the professional development of younger engineers

In submitting an amplified record of practical engineering experience, the applicant must be sure to define clearly responsibilities in the different engagements. These are best shown by specific examples which give the extent of the applicant's relation to the engineering work. It does not suffice to say that one was in charge of the design of a new tool or manufacturing process. The board must have complete details. If possible, one should show what new features were added and indicate length of this particular engagement.

YOUR ENGINEERING EXPERIENCE JOURNAL—FIRST STEP TO PE REGISTRATION

Before you can become a registered professional engineer, you have to show that you have had adequate experience. Here's how to document it by keeping a journal, to simplify this task.

When you file your application to become a registered professional engineer, not only must you have the proper experience of a grade and character acceptable to your board of registration, but you must also document this experience in a manner that the board can evaluate, verify, and assign full credit for.

The purpose here is to outline a method of keeping a journal of your experience which will simplify this all-important task.

Boards find that most engineers are too modest in documenting their hard-earned experience on the application form, thereby not realizing full effective value. On the other hand, the applicant should not boast needlessly.

In evaluating your engineering experience, the board of examiners must reach a unanimous decision based on what is written into the record. Board members cannot read between the lines and they cannot go by mere hearsay. It is not enough to list job titles, no matter how impressive. The written record must describe your duties and responsibilities in detail, leaving no doubt as to the nature and extent of your engineering experience.

The experience factor is the most variable among the requirements of boards of examiners. Most registration laws require a minimum of 4 years of experience in engineering. Details of such requirements may be obtained from the boards directly.

What Is Satisfactory Experience?

There is no one clear-cut guide to determine what is satisfactory and what is not. Generally, however, boards look for a pattern of progressively responsible design experience. Experience must be of a nature and character approved by the board and should be broad in scope. Not merely number of years but the nature of such experience is the important ingredient.

The engineer should seek the widest possible responsibility. Experience should be progressive and of increased standard of quality, responsibility, and initiative. As for contracting, military, and teaching experience, refer to the pertinent sections of this book. An up-to-date definition of "responsible charge" is also included in the Appendix.

The Journal

The registration-minded engineer should plan ahead and not trust to memory when the time comes to document experience. A recommended journal outline will contain at least the following information with respect to each project worked. Each completed project-description form should be filed away chronologically for future reference.

Journal Outline and Typical Journal Entry

Name of project
Duration of project
Engineering supervisor (registered how, where, and when)
Description of project
Specific duties performed
Degree of responsibility
Engineering decisions rendered

An entry in the journal might appear as in the sample that follows. With this kind of documentation of progressively responsible engineering experience, your board will certainly be able to make an evaluation properly and completely.

SAMPLE JOURNAL ENTRY

Name of project:
Louisville Acetylene Plant
Duration of project:
August 1982 to December 1985
Engineering supervisor:
Thomas Henry, PE; registered New York State by written examination (1952).
Description of project:
Develop method of acetylene generation continuously by wet process using calcium carbide.
Specification of duties performed:
Given a rate of gas generation, carbide and water rates were calculated, using the basic chemical equation. Sludge effluent to settling pond estimated and

sludge pumping system was designed. A battery of 20 similar generators were designed and specified using vibrating feeders.

Degree of responsibility:

Under direction but not complete supervision of Mr. Henry.

Engineering decisions rendered:

Finalized water and carbide rates; ran operating tests on vibratory feeders; finalized generator sizes and volumes.

PROBLEMS IN EVALUATION OF EXPERIENCE RECORDS

A continuing concern of boards of registration has to do with evaluation and interpretation of experience records of applicants for licensure. Putting it mildly, many of the applications received leave much to be desired, particularly in the use of vague terminology to describe the applicant's specific duties and degree of responsibility in each engagement and level of employment.

It is quite a common practice by applicants to say something like, "My duties consisted of designing power plants, including economic evaluation of the sites and original investment costs, etc., etc.," a paraphrasing of the definition of mechanical engineering as given in the statutes. Very often, very general phrasing is used, such as, "I was engaged in . . . ," or "I assisted in . . . ," or "I performed . . . ," or "I worked on the job site with the construction crew," but no mention whatsoever is made as to specific work duties and specific responsibility. The boards cannot tell from the descriptions whether the applicant was a draftsperson, designer, print coordinator, or none of these—or perhaps all of them at varying times.

In other instances, a statement by an applicant who may be a high official in government might say, "I had the full responsibility for the following projects, etc., etc.," without mentioning in what capacity or specific relationship. Many upper-level administrators who have no knowledge of or ability in mechanical engineering whatsoever but who oversee departments responsible for mechanical engineering activities can make the same statement. Titles are no criteria!

On other occasions, an applicant may clearly indicate the specific level of responsibilities and duties, but deliberately cloak the fact that, in a shown span of time, those duties and responsibilities occurred only on a part-time or occasional basis. Perhaps the experience record may be written up so that many other duties or gainfully employed activities not related to mechanical, civil, chemical, electrical, etc., engineering are deliberately omitted, making it appear that all the indicated time span was devoted to qualifying experience. Here the applicant would be advised to state those proportions of time devoted to qualifying as acceptable experience or other forms of experience.

The applicant should be strongly encouraged to carefully avoid omissions or the use of vague language and general terminology in describing work responsibilities. If the applicant feels that such vague and general terminology is necessary to cloak or hide deficiencies or shortcomings, then that individual knows that fully qualifying experience has not transpired. As a suggestion, an applicant should first prepare a "working draft" of the experience record, and then study it in the light of the above foregoing discussion.

If language has been used that is so general that a board cannot determine the applicant's specific duties (all of them) and specific level of responsibility and specific duration of each such level of responsibility, then the experience record should be redrafted to be fully precise and clear, leaving no doubt in the board's mind as to the full value of that experience.

Preparing the Application

The applicant can expedite action on an application by the care with which it is prepared. All information on the form, with the exception of signatures, should be either typewritten or printed. If it is complete, correspondence with the applicant is obviated. If the experience and degree of responsibility are described in sufficient detail, further questions will be unnecessary. If the references respond promptly, delay will be avoided. If the application and reference forms are typewritten or printed legibly, a more favorable impression will be created.

From the outset, the applicant will be seeking to make a favorable impression. The applicant will wisely attempt to make the best possible impression on the members of the board of examiners.

Application Photograph

All boards will require some means of identification of the candidate. Usually two photographs in full face are necessary, one to attach to the application form and one for purposes of admission to the examination room. Photographs should be large enough so that the width of the subject's face is not smaller than ¾ inch.

Application Form

The board of examiners supplies the application forms and associated papers to be filled out by the applicant. The forms contain statements to be made under oath, showing the applicant's education and detailed summary of technical work, and contains not less than five references, of whom three or more shall be registered professional engineers having personal knowledge of the applicant's engineering experience.

The registration fee is indicated for all conditions of licensure.

In some states, one application form is used for both EIT classification and full PE licensure. In other states, two forms are used, that for EIT being more simplified than the full PE form. Full instructions are included with the application. When writing for the application, the applicant should request a copy of the registration statute.

Processing the Application

How do boards go about processing the applications? As one example, let us see how the New York board operates.

The Bureau of Professional Licensing Services in Albany is the central agency for the New York State Education Department for maintaining all permanent files. The Bureau dispenses application forms and other materials; certifies matters of fact and record, such as age, citizenship, fees, and education evaluations; administers the written examinations; and processes licenses and registrations. It also is responsible for collecting references mailed in by sponsors for the applicants. All records are kept on file in Albany, and all correspondence should make reference to the file.

All records are kept on file forever and only other boards may request information therefrom through proper channels.

More than half of the qualification requirements are matters of fact and do not require the considered judgment of the board. Like other states, New York has seven qualifications (see Chapter 6):

1. Age
2. Citizenship
3. Good moral character
4. High school graduation
5. College education or equivalent
6. Experience internship
7. Professional examinations

It doesn't take board action to determine that an applicant is at least 19 or 25 years of age, as the requirement may be. This is a matter of record, just as are citizenship and high school and college graduation.

When an application is requested, a complete set of forms is mailed. This set consists of:

1. Instruction sheet
2. Record card
3. Application form
4. Professional Certification Form 2E
5. Reference forms (five)
6. Self-addressed envelope

When the filled-out application is received at Albany, it enters the system at the Bureau of Accounts. If a check for the fee is not included a bill is sent to the applicant, and the application is placed in a temporary file. Upon payment of the fee, the application is sent to the Bureau of Professional Licensing Services. As soon as this bureau verifies such matters as stated age of applicant, etc., and the required reference forms or letters have been received, the entire file of the applicant is sent to the secretary of the board for consideration.

The applicant has been instructed to give one reference form to each of five references or character and experience sponsors. Although these are returned directly to the secretary, the Bureau of Professional Licensing Services assembles them with the application papers before clearing them for board consideration.

The board reviews the matters certified by the bureau. In addition, considerable attention is paid to the experience record and what the applicant's sponsors say about the applicant. Then the board determines what lawful disposition can be made of the case. This is accomplished in the form of a recommendation to the state education department.

Official departmental action comes only as the result of board recommendation.

The board meets twelve times a year. It takes action on over 3000 applications, most of which are new. Each application, all supporting documents, and all related correspondence are enclosed in an envelope bearing the applicant's name. A number of cases are sent to the secretary at one time, the number depending upon the current activity. A record is kept of each and every application, and a work sheet is added to each envelope. The secretary then reviews each complete individual file and suggests what the board action on it should be by making a notation on the work sheet.

The secretary's staff then assembles the applications into groups of fifteen or so and sends the bundles to board members, in order that each of them may study all cases before formal board action takes place. This procedure ensures the studied consideration of all board members. *All opinions are carefully weighed, and the final decision must be agreed upon unanimously.*

The Bureau of Professional Licensing Services notifies the applicant by mail. If the applicant does not qualify for the written examination, he or she may be called in for an interview to clarify his or her position. The applicant with outstanding professional experience, in addition to graduation from an accredited engineering program, may receive notification of licensure by endorsement without written examination. This is extremely rare.

This latter method of licensing (i.e., without written examination) may lead to difficulties in getting licensed in other states. To receive a license in other states, one must meet the statutory requirements of that state at the time of original licensure. At this time all states require 16 hours of written examination; therefore, someone obtaining licensure without a written exam would probably be required to sit for an exam in another state before being granted comity.

10

THE WRITTEN EXAMINATION

When we discussed the FE examination in Chapter 5, the importance of mathematics and the basic sciences were stressed. We showed that in the majority of cases, the written examination, for the present, is the only way that separates the competent from the incompetent in the eyes of the law. We also saw that the examination must be a sampling process, with a wide distribution of samples.

SCOPE OF THE 16-HOUR EXAMINATION

The examination for professional engineer is divided into two sections: Fundamentals of Engineering (FE), Part A, and Principles and Practice of Engineering (PE), Part B. Samples of questions from the FE examination are shown in Fig. 10.1. There are two 4-hour sections for each part, for a total of 16 hours of written examinations. All states administer examinations that are 16 hours long.

Fundamentals of Engineering (FE)—Part A (A.M.)

The morning part of the FE exam consists of machine-scored multiple-choice questions, each with five possible responses, four of which are what are called "distractors." All parts of the examination are open-book; that is, the examinee may use textbooks, handbooks, bound reference books, and a battery-operated silent calculator or computer. No writing tablets, unbound tablets, or unbound notes are permitted in the exami-

Considering the commitment of the U. S. Congress and many technical societies along with the trend in engineering schools to move toward voluntary use of the metric system, the Uniform Examinations and Qualifications for Professional Engineers (UEQ) of NCEE has recommended and the Board of Directors has approved the formulation of appropriate questions in both SI units (System International) and non-SI units for use in the Fundamentals of Engineering Examination.

In the Morning Section, problems that require the manipulation of numbers and units are given in SI units and non-SI units. Each problem statement requires the same reasoning process; you may answer either problem statement for equal credit. Please see the SI and non-SI problem statements of a sample problem below.

Select either problem

1. (SI) m_1 = 1 kilogram m_2 = 2 kilograms

Before Impact: v_1 = 2 m/sec→ v_2 = 1 m/sec←

After impact: v_1 = 1 m/sec← v_2 =

Two objects are on a head-on collision course. The masses of the objects, their velocities before impact, and the velocity of object 1 after impact are shown above. What is the velocity of object 2 after impact?

(A) 0.5 m/sec→ (B) 0.5 m/sec←
(C) 1.0 m/sec→ (D) 2.5 m/sec→
(E) 2.5 m/sec←

1. (non-SI) m_1 = 1 lbm m_2 = 2 lbm

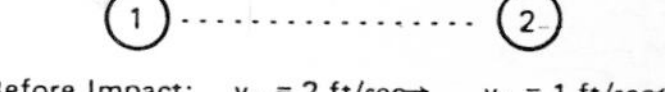

Before Impact: v_1 = 2 ft/sec→ v_2 = 1 ft/sec←

After Impact: v_1 = 1 ft/sec← v_2 =

Two objects are on a head-on collision course. The masses of the objects, their velocities before impact, and the velocity of object 1 after impact are shown above. What is the velocity of object 2 after impact?

(A) 0.5 ft/sec→ (B) 0.5 ft/sec←
(C) 1.0 ft/sec→ (D) 2.5 ft/sec→
(E) 2.5 ft/sec←

In the Afternoon Section, numbers and units are given in both SI and non-SI units. A number and unit in one system is followed by the equivalent number and unit in the other system. Please see the sample problems below.

Questions 1-2 relate to the air-flow system shown below.

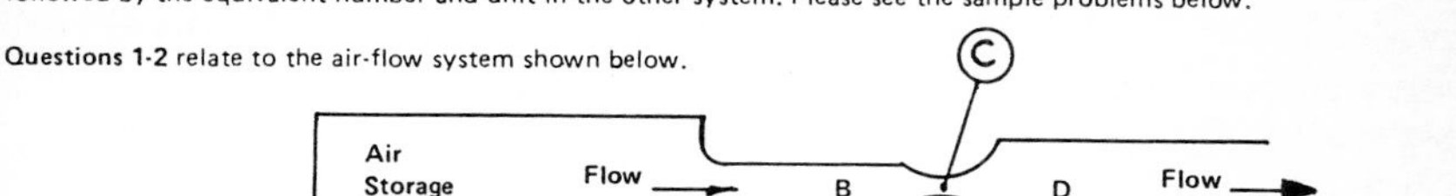

All pressures are absolute.
Pressure at A = 100 lbf/in^2 (6.9 x 10^5 N/m^2)
Temperature at A = 1000° R (555° K)
Pressure at B = 90 lbf/in^2 (6.2 x 10^5 N/m^2)
Throat area at C = 0.50 ft^2 (4.6 x 10^{-2} m^2)

1. The mass density of the air inside the tank A is most nearly

(A) 1.88 x 10^{-3} lbm/ft^3 (3.01 x 10^{-2} kg/m^3) (B) 7.52 x 10^{-3} lbm/ft^3 (0.120 kg/m^3)
(C) 0.270 lbm/ft^3 (4.33 kg/m^3) (D) 0.357 lbm/ft^3 (5.72 kg/m^3)
(E) 0.532 lbm/ft^3 (8.52 kg/m^3)

2. The speed of sound inside the tank A is most nearly

(A) 760 ft/sec (232 m/sec) (B) 1120 ft/sec (342 m/sec)
(C) 1310 ft/sec (400 m/sec) (D) 1550 ft/sec (473 m/sec)
(E) 1820 ft/sec (555 m/sec)

Figure 10.1 (SOURCE: NCEE.)

nation room. Examinees are not permitted to exchange any reference materials or aids during the examination.

Both A.M. and P.M. sections of the FE examination must be taken to receive a score. Each of the 140 multiple-choice questions from the subjects listed in Table 10.1 has five answers from which the one best answer must be selected. Answer every question; guess if necessary. You are encouraged to guess: your score depends on the number of questions answered correctly and no subtractions are made for incorrect answers, so it is to your advantage to guess rather than leave a question unanswered. Some questions are presented in both English and metric units and the examinee may work either question for equal credit.

Fundamentals of Engineering (FE)— Part A (P.M.)

This part of the examination is also machine-scored and consists of problem sets. (See Table 10.1.) The test consists of a total of seventy questions: fifty required questions, plus a choice of a total of twenty questions from any two of the five additional subjects offered. Answer the questions in the four required subjects (the first fifty questions). Then choose two subjects from the five additional subjects and answer the ten questions in each. The approximate number of questions in each major subject is shown in Table 10.1. As the practice of engineering education changes, the format and subject matter of the examination may follow suit.

Principles and Practice of Engineering (PE)— Part B (A.M. and P.M.)

After the FE examination has tested the examinee's facility in mathematics and engineering theory, the PE questions require the candidate to show proper judgment in selecting correct formulas, economical considerations, and practical approaches. Problem wording and terminology are so arranged that the examinee must call upon professional experience to develop a solution.

This part of the examination is professional in nature and designed for persons with several years of experience, involving the application of scientific principles to the problems the engineer meets in everyday work, in the engineering office and in the field.

Part B consists of two four-hour sections, one in the morning and one in the afternoon. This part of the examination is not multiple choice but is of the subjective type. Textbook problems, proofs, and derivations are

TABLE 10.1

Morning Section of Part A	
Subject	**Approximate percent**
Chemistry	7%
Computer Programming	6%
Dynamics	12%
Electrical Circuits	13%
Engineering Economics	4%
Fluid Mechanics	10%
Materials Science	4%
Mathematics	12%
Mechanics of Materials	9%
Statics	9%
Structure of Matter	4%
Thermodynamics	10%
Total A.M. Questions = 140	100%

Afternoon Section of Part A*	
Required subjects	**No. questions/subject**
1. Engineering Mechanics	15
2. Mathematics	15
3. Electrical Circuits	10
4. Engineering Economics	10
Additional subjects (Choose 2)	**No. questions/subject**
5. Computer Programming	10
6. Electronics and Electrical Machinery	10
7. Fluid Mechanics	10
8. Mechanics of Materials	10
9. Thermodynamics/Heat Transfer	10
Total P.M. Questions	100

* Only 70 questions are to be answered. SOURCE: NCEE

avoided. These sections are intended to show the examinee's ability to apply sound engineering principles and judgment to the solution of problems normally encountered in practice. This examination is hand-graded, not machine-graded, for the time being (there are forces afoot to make it multiple choice). In each section, the examinee is required to respond to four problems. There is one economics problem in the afternoon section only.

Only the most general engineering principles are required for the solution of the problems. Considered important are the application of good engineering judgment in the selection of and the evaluation of pertinent information and the ability to make reasonable assumptions when necessary. Answers submitted for some problems may vary according to assumptions made. Partial credit may be given if correct fundamental engineering principles are applied, even though the final answers may be incorrect. This part of the examination is hand-graded by professional engineers. This part is also open-book and completely in the English system of units.

For information regarding specific requirements in your state and the guidelines to your state's administration of the NCEE uniform examination, contact your board's office. Addresses of state boards may be found in the Appendix. Typical sample questions on the examinations and further information on details relating to the formats currently applicable may be obtained by writing NCEE.

The administration of each state's examination is controlled by the state board of registration. Your board will provide you with complete instructions as to which disciplines in Group I (required questions) and Group II (additional subjects) will be offered for your specific examination. Be alert to the fact that you should consider problem groups that will provide you with a broad selection. For instance, select problems in all groupings; if you are a mechanical engineer, select groupings that contain problems in heat transfer, fluid flow, and pumping in the appropriate discipline, i.e., chemical engineering, mechanical engineering, civil engineering. See Table 10.2.

HOW YOU CAN PASS THE FE EXAMINATION—A CASE HISTORY

Here is an actual case history of an electrical engineer who did an excellent, well-planned job of thoroughly preparing to sit for the FE examination: how he proceeded, what references he used, what he took into the exam room, and his unique approach that led to his successful bid. The author hopes that his story will fire your imagination and lead to your own successful attempt, whether or not you are an electrical engineer, as he was.

How He Prepared

As a starter, he chose one basic reference book in engineering fundamentals and used it to begin to review his strong points (mathematics, electrical theory, physics) and to relearn his weaker subjects (thermodynamics, economics, nucleonics). By limiting himself to one such reference book, he was able to familiarize himself with the types of problems

TABLE 10.2 Principles and Practice of Engineering Examination—Examination Specifications Effective with the Spring 1983 Administration

Group I Examinations Administered in Spring and Fall			
Discipline	Category	Subject	Approximate no. problems/ category
Chemical	1	Thermodynamics	3
	2	Process Design	4
	3	Mass Transfer	4
	4	Heat Transfer	2
	5	Chemical Kinetics	4
	6	Fluids	2
	7	Engineering Economics	1
		Total	20
Civil/Sanitary/ Structural	1	Transportation	5
	2	Structural	7
	3	Sanitary	5
	4	Hydraulics	4
	5	Soils	2
	6	Engineering Economics	1
		Total	24
Electrical	1	Power	8
	2	Electronics	6
	3	Control Systems	5
	4	Computers	4
	5	Engineering Economics	1
		Total	24
Mechanical	1	Mechanical Design	8
	2	Management	1
	3	Energy Systems	6
	4	Control Systems	1
	5	Thermal and Fluid Processes	3
	6	Engineering Economics	1
		Total	20
Group II Examinations Administered in Fall Only			
Aeronautical/ Aerospace	1	Structures	7
	2	Aerodynamics	4
	3	Flight Mechanics	4
	4	Propulsion	4
	5	Engineering Economics	1
		Total	20

TABLE 10.2 (*Continued*)

Group I Examinations Administered in Spring and Fall			
Discipline	Category	Subject	Approximate no. problems/ category
Agricultural	1	Irrigation & Drainage	3
	2	Soil & Water Conservation	3
	3	Machine Design	2
	4	Power, Control & Systems	2
	5	Structures	2
	6	Environmental	2
	7	Crop Handling & Processing	3
	8	Food Engineering	2
	9	Engineering Economics	1
		Total	20
Ceramic	1	Phase, Microstructure, Mechanical & Electrical Properties	4
	2	Processing	4
	3	Glass Technology	2
	4	Refractories, Corrosion, Heat Transfer	3
	5	White Ware, Glazes, Enamels, Coatings	2
	6	Structural Clay, Plant Operation, Plant Design, Mineral	4
	7	Engineering Economics	1
		Total	20
Fire Protection	1	Hydraulics	4
	2	Suppression systems	2
	3	Fire Behavior	2
	4	Fire Communications	2
	5	Hazards	4
	6	Structures	4
	7	Fire Protection Management	1
	8	Engineering Economics	1
		Total	20
Industrial	1	Facilities Planning	3
	2	Management	2
	3	Manufacturing	3
	4	Operations Research/Systems	3
	5	Production & Inventory Control	3
	6	Quality Assurance & Applied Statistics	2

TABLE 10.2 (*Continued*)

Group I Examinations Administered in Spring and Fall			
Discipline	Category	Subject	Approximate no. problems/ category
	7	Work Measure & Ergonomics	3
	8	Engineering Economics	1
		Total	20
Manufacturing	1	Processes	4
	2	Operations	3
	3	Production Facilities	6
	4	Manufacturing Management	4
	5	Systems	2
	6	Engineering Economics	1
		Total	20
Metallurgical	1	Fabrication & Processing	2
	2	Metallurgical Thermodynamics and Corrosion	3
	3	Selection of Material	2
	4	Thermal Treatment & Solid State Processing	3
	5	Structure/Property Relationships	2
	6	Failure Analysis	2
	7	Mineral Processing	2
	8	Extractive Metallurgy	3
	9	Engineering Economics	1
		Total	20
Mining/Mineral	1	Exploration & Geology	3
	2	Mine Planning	6
	3	Mine Operations	3
	4	Ground Control	3
	5	Mineral Processing	3
	6	Environment & Governmental Regulations	1
	7	Engineering Economics	1
		Total	20
Nuclear	1	Thermal Power Plant Operation	4
	2	Radiation Protection & Instrumentation	3
	3	Neutronics & Core Physics	2
	4	Reactor Engineering	2
	5	Nuclear Fuel Cycle Analysis	2
	6	Safety and Licensing	2

TABLE 10.2 (*Continued*)

Group I Examinations Administered in Spring and Fall			
Discipline	Category	Subject	Approximate no. problems/ category
	7	Shielding	2
	8	Waste Management	2
	9	Engineering Economics	1
		Total	20
Petroleum	1	Drilling and Completion	6
	2	Production	4
	3	Reservoir & Improved Recovery	6
	4	Formation Evaluation	3
	5	Engineering Economics	1
		Total	20

NOTE: Above listing was obtained from the *Registration Bulletin*, December 1982, courtesy the National Council of Engineering Examiners. Table reflects the findings of NCEE in their Task Analysis Survey to determine specific task areas that licensed professional engineers perform on the job.

treated, and of equal importance, where in the book they were to be found. This approach he found quite helpful during the course of the examination.

For a listing of reference books, see Recommended Reference Texts and Study Aids in the Appendix.

In further preparation, he made use of his class notes from the basic engineering science review course he attended, which had been provided by his company.

Next, he incorporated all of the important formulas, equations, principles, etc., from his sources into a cloth-bound composition pad which he sectioned off by subject area for quick reference, using tabs to identify the subject areas.

During this part of his preparation, he gave little attention to totally unfamiliar subjects (fluid mechanics, strength of materials).

What He Took to the Examination

His basic engineering reference book on fundamentals

His cloth-bound composition pad

Tables of conversion factors:
English-to-metric conversion factors
Trigonometric functions

His college thermodynamics textbook containing examples on the Otto cycle, closed boundary problems, and steam tables

Battery-operated, silent, nonprinting calculator, slide rule for backup, eraser, pencils, ballpoint pens, and extra batteries for his calculator

Test Format

Morning section: 140 multiple-choice questions in twelve hard-to-distinguish subjects. (See Table 10.1.)

Afternoon Section: Seventy questions in all. (See Table 10.1.) Each problem consists of four sets of ten or fifteen questions; choose two groups from additional subjects.

Examinee's Technique

Morning section. He proceeded to answer all those questions related to electrical theory. Then he repeated the same procedure for the next subject in which he felt confident. Thus, for the first hour very little guessing was used; however, as he attempted those problem types of increasing difficulty, he passed over those questions that were unfamiliar to him. At times he would skip entire problem groupings due to lack of familiarity. By applying this technique, he was able to cover or at least see all 140 questions within the 4-hour limit.

Then, 15 minutes before the end of the 4-hour time limit, he went through the entire sheet and filled in all the unanswered questions with predetermined "guess letters" which appeared to be reasonable and correct. Note: Of the 140 questions, if 70 to 80 of them are answered correctly, the examinee will have passed that part.

Afternoon section. He used the same approach, attending to his strong questions first. Before starting any problem, he would scan the entire problem set to ensure that all ten or fifteen questions within the problem set were uniformly familiar. If so, he would then start solving. If not, he looked elsewhere. He had to keep in mind that no credit would be given for answered questions in problem sets beyond the four required problem sets he had chosen to answer. Again, the one-letter guess was employed 15 minutes before the time limit expired on all unanswered questions in his four selected problem sets.

Comments

Since no scrap paper is allowed (and there is very limited scrap space available in the question booklet) he used his bound composition pad for calculations.

If you use a calculator, remember to turn off the unit when not in use.

Be prepared for a 15-minute lunch break, due to either a late start in the morning or long waiting lines at the food service available, since everyone will be finished with the first part of the exam at the same time.

Watch for poor wording of the question group which can lead you to the most reasonable answer with little thinking effort. A question that is worded poorly can guide you to which of the options might be most correct. You could get credit for answers just on the basis of a random guess, and most examinees guess, as we saw above.

OBJECTIVE-TYPE EXAMINATIONS

The Fundamentals of Engineering (FE) Part A morning and afternoon sections are now objective-type exams. The purpose of using this type is to probe for a broad understanding of the fundamentals of engineering. Both sections are multiple choice and are given in both English and metric systems, where appropriate.

An objective-type question must be carefully prepared to serve its purpose. Each multiple-choice question is intended to test the examinee's analytical ability as well as knowledge of a given engineering principle. First the problem is stated and then five answer choices are listed, one correct and the remaining four incorrect (distractors). The incorrect answers have been selected to seem logical to one who does not completely comprehend the principles involved in the question.

By using this type of question, many more fundamental engineering principles can be covered in a given time than with the subjective-type questions used previously. Boards can have a better measure of the examinee's understanding and use of sound engineering principles since many more questions can be answered. Note that these multiple-choice questions are a far cry from the old true-false questions which were introduced many years ago. The sophisticated tests used today do give a much more accurate measure of an examinee's proficiency. It takes much longer to prepare an objective-type question than it takes to prepare one of the subjective type; however, the grading is much faster and much more consistent. It eliminates differences between graders and differences in any particular grader because the tests can be machine-scored.

The change from a subjective- to an objective-type examination has come about due to the pressure of numbers and the recognition that such a technique is capable of making a positive contribution to the quality of the examination.

Insofar as the acceptance of objective testing techniques is necessitated by the numbers of candidates sitting for the exam, it is analogous to the acceptance of an unpalatable but necessary medicinal dose by a small child—it was resisted to the bitter end and was finally ingested with bad grace. More than that, the resort to objective testing was by some accompanied by a sense of defeat and a bad conscience. Standards, it was felt, had been compromised and the best that can be said is that the problem of numbers left no other choice.

Much better, it seems to me, would be to consider objective testing techniques for what they are worth rather than merely as a stop-gap or a distasteful compromise. What positive benefits do they have to offer which, in combination with the traditional techniques, might produce a stronger examination than either approach used to the exclusion of the other? Seen in this light, the question of the number of candidates to be tested becomes less relevant to the decision about the use of objective testing. An examining board which is not yet faced with the problem of numbers might find justification for including objective material solely on the grounds of improving the quality of its examination. It is this possibility, namely that objective examination techniques have much to recommend them in their own right, that should be explored in greater depth.

Essay- versus Objective-Type Examinations

What are the strong points of the essay-type question? Perhaps their outstanding advantage, one that is matched by a corresponding deficiency in objective examinations, is that in an essay examinees must produce and organize ideas on their own. Possible answers are not supplied for them and the process of arriving at a right answer in not based on recognition, as is usually the case in objective examinations. In defense of objective examinations, it should be remembered that the two abilities—to recognize right answers and to produce them in the absence of stated choices—are probably highly correlated.

Second, essay answers provide an opportunity to observe the examinee's competence at producing an extended composition which is acceptable as writing, that is, on grounds of style and correctness. Third, an essay examination permits the display of the intellectual processes leading to a conclusion. On occasion, it may be important to see that an examinee has gone about the solution to a problem in a sensible and professional way, even though the answer arrived at is wrong in some particulars.

Advantages of Objective Methods

Now let us consider the advantages of objective methods of examination. First, there is the matter of comprehensiveness of coverage. Because examinees do not supply the answers through the time-consuming processes of writing, they can be made through objective questions to consider many more aspects of a subject in a given amount of time than is possible through essay questions. This would mean that an objective examination is likely to provide a fairer estimate of the examinee's knowledge of the field as a whole. From the examinee's point of view, comprehensiveness of coverage means less likelihood of being downgraded by getting one or two essay topics in areas where he or she happens to be weak, even though he or she may be reasonably competent in the field as a whole.

Second, objective questions are more effective in forcing candidates to face problems and issues to which the examiners want candidates to address themselves. It is well known that skillful and fluent writers can, in essay questions, make themselves look much better than they are by reshaping the question to their own purpose.

Third, grades on objective examinations are more likely to be reliable than those on essay examinations, where multiple grades may have varying standards. Reliability is a statistically measured concept referring to the extent to which an observed grade for a given candidate corresponds to the grade which most truly represents the candidate's ability.

Fourth, a positive characteristic of objective examinations is the ease with which they can be graded. The use of high-speed data processing equipment makes the actual scoring of tests in the quantities likely to be generated a relatively easy task. It should be remembered that objective examinations are not at all easy to produce, however. In an essay examination, it is relatively easy to produce the problem cases, while it is very difficult and time-consuming to grade the resulting essays. The reverse is true of objective examinations. The time, effort, and creative thinking go into producing the examination, while the grading is largely a mechanical process.

Thus it is probably true that, overall, no less effort is required of the board to produce objective examinations than is required to produce essay examinations. If a decision is to be made in favor of objective examinations, it should therefore be based on other considerations, namely their greater fairness to examinees, their greater amenability to control by the examiners, and the greater reliability possible in scoring them. Also, with regard to economy of effort, while objective examinations do require time in making up, the demands they make are probably timed better. The grading procedures are likely to be carried out under great pressure and it is exactly then that objective examinations make speed possible. On the other hand, while objective examinations are difficult to write, it is just at this stage that time is available, assuming careful organization of the writing procedure.

I should like to spend some time responding to a frequently adduced criticism, namely that such examinations operate at the level of triviality; that is, they are capable of measuring only memory for fact and the results of rote learning. Conversely, they cannot measure the higher intellectual processes such as logical reasoning, analysis, and the application of knowledge. Such opinions seem to betray ignorance of the broader potentialities of multiple-choice questions. To attack the matter at its simplest level, objective tests should not be equated with true-false or yes-no questions. These are relatively rudimentary forms of multiple-choice questions and their possibilities are certainly limited. Competencies which can be satisfactorily measured through these methods might tend to be of low-level order—perhaps, as some have said, they may be "trivial." But multiple-choice questions can be written in far more sophisticated forms. For one thing, the examinee can be faced with more possible answers (four or five are fairly common) so that he or she must consider more aspects of the question in arriving at an answer. Moreover, instead of posing possible answers in such categorical and simplistic terms as yes-no, true-false, possible answers of greater complexity requiring a deeper understanding of the question and a greater ability to draw fine but significant distinctions can be formulated.

Advantages of Machine Scoring Available with Objective Examinations

One of the most important advantages of machine scoring is its speed. Several thousand exams can be scored in a matter of a day or two. A second advantage is the low cost for machine scoring. Since human labor involved is limited to a minimum amount, the costs are quite a bit lower (after the initial cost of hardware) to score exams. A third advantage is that there is *equal* scoring for all examinees. The scanner or computer that performs the scoring is programmed to perform the scoring in the same manner for each test, so that one can be assured that every examinee is scored in exactly the same way for each and every test. The possibility of errors being introduced by hand scoring do not occur, so increased test reliability occurs.

A fourth advantage is legal security. If every exam is scored in exactly the same way through the operation of a machine, there is less chance that anyone would challenge the results on the basis of a particular exam being scored unfairly.

Greater test security is another advantage. Since the exams are scored quickly and all answers are recorded in the machine, there is less probability that the recorded data will fall into the hands of anyone not entitled to the information. There is greater control over the scores, the test items, and the answers themselves.

On the other hand, machine scoring is not the last word. There are several disadvantages that are frequently tied to machine scoring. First, machine scoring does not permit open responses. This can become a problem in that a person may know the answer to a particular problem, but may not find an answer in the choices given which matches her or his answer. Essay format permits creativity so that the examinee can express an answer in several ways. It is difficult to give the partial credit through the medium of machine scoring that is possible with essay format.

Since multiple-choice questions are based on the right or wrong basis for the answer, the examinee does not get half credit or choosing one that is fairly close to the right answer. The examinee has either the right answer or the wrong one. With hand scoring, it is possible to look at the examinee's answer and give partial credit for an answer that reflects some knowledge in the way the answer is developed, even though it is incorrect.

Another possible problem is that if a test question is not properly developed, a poorly worded question can more or less guide the examinee to the option that is most correct, even if the examinee did not have the necessary knowledge to answer the question. Related to this is a very important point, and it is in regard to the matter of permissive and encouraged guessing as given in the NCEE instructions. The examinee can get credit for answers just on the basis of a random guess.

Finally, a possible but distinct disadvantage of machine grading is that it may force problems to be divided into small "testable" pieces rather than allowing problems with large scope, requiring a great deal of conceptual work. Instead, the problem is divided into small segments so that the examinee can be guided through examining each segment of her or his knowledge. In this procedure, the examinee's overall grasp of material and concepts may not be tested.

Since these advantages and disadvantages apply to any examination,

NCEE is exploring the trade-offs involved between hand scoring and machine scoring and has authorized a feasibility study.

Preparation of Objective-Type Questions

An objective-type question must be carefully prepared to serve its purpose. The problem is stated and then four or five answers are listed, one right and the others wrong, or vice versa. The wrong ones include some which would be obtained if a wrong assumption were made, or if some important information was not included in the solution. The incorrect choices have been selected to appear to be logical to an examinee who does not completely understand the principle being tested. Here is an illustration:

- A fan delivers air to the inlet of a ventilation duct at 2000 ft^3/min and 0.8 in Hg static pressure. The inlet to the fan is at 0.2 in Hg vacuum due to filters in the inlet line. Fan speed is 1200 r/min and the fan shaft input is 0.4 hp. Duct air velocity is 800 ft/min. It is desired to obtain an air flow of 4000 ft^3/min by increasing the fan speed without making any changes in the mechanical adjustments of the vent duct. Which of the following statements regarding the new operations is *not* correct?

  (a) The new fan static pressure at discharge will be 3.2 in Hg
  (b) New fan speed will be 2400 r/min.
  (c) New fan shaft input will be 1.6 hp.
  (d) New air duct velocity will be 1600 ft/min.
  (e) The fan inlet will have 0.8 in Hg vacuum.

The examinee must understand the principles of a centrifugal fan, the ratios between output, speed, pressures, and power, whether directly as the square, or as the cube of other functions. If you picked (c) as the only incorrect answer, you understand the principles of fan performance.

Another type of objective question is a root or base statement about which mathematical or descriptive answers are required for ten independent questions or subsets. The answers required for each of these subsets are only arrived at by properly solving the problem by a manual process which requires judgmental and mathematical engineering ability.

The subsets are independent, even though at first glance it might appear that an answer to an earlier subset is being reused to answer a later one. If the examinee fails to answer a previous subset, she or he can still by proper application of engineering theory and judgment calculate all elements required to respond to any subsequent subset.

Each subset has a choice of five answers. One is absolutely correct; the four other answers appear rational if a normal error in formula, mathematics, or interchange of engineering units is made. The incorrect answers (distractors) are calculated by a consultant who prepares the problem with the same diligence used in preparing the base statement, the subset questions, and the correct answer. An illustration follows:

- As the plant engineer assigned to study air pollution problems in your plant, you are to determine the flow rate of air in the duct leading to a large blower. Not having air flow instruments, you decide to do this by adding a small amount of sulfur dioxide to the airstream just upstream of the blower suction and then analyzing a sample of the airstream leaving the blower discharge for sulfur dioxide. You assume that the blower will provide good mixing, that the sulfur dioxide does not react chemically with the airstream, and that the air is dry. The basic operating conditions for the blower are: at the entrance to the blower, absolute pressure is 14.0 lb/in^2 at 78°F; at the blower exit absolute pressure is 15.0 lb/in^2 at 80°F.

PROBLEMS

1. If the sulfur dioxide is added at the rate of 90 ft^3/h, at 18.0 lb/in^2 (absolute) and 60°F, the addition rate of the sulfur dioxide in sft^3/h is:
(a) 70 (b) 90 (c) 98 (d) 104 (e) 116

2. If a sample of the blower exit gas (air plus sulfur dioxide) is taken at a temperature 80°F and has a volume of 250 mL at 745 mmHg, the gram mols of mixture in the sample are:
(a) 0.00932 (b) 0.00997 (c) 0.01116 (d) 0.01201 (e) 0.01250

3. An analysis of the sample reports that it required 6.5 mL of 0.0020 *N* base to titrate; therefore, the g of SO_2 in the sample was:
(a) 6.5×10^{-6} (b) 4.16×10^{-4} (c) 8.32×10^{-4} (d) 1.16×10^{-3} (e) 8.32×10^{-1}

4. If the plant laboratory reported a sulfur dioxide concentration of 0.0060 millimoles per liter of sample (at S.C.), the concentration in ppm (by volume), would be:
(a) 6 (b) 134 (c) 269 (d) 61,000 (e) 134,000

5. In a series of tests, sulfur dioxide was added at the blower exit at a rate of 20 lb/h. The laboratory reports that the sample you sent them had a volume of

2.65 liters, measured at 750 mmHg and 72°F, and contained 150 *ppm* SO_2 by volume. The SO_2 addition rate, in sft^3/min, was:
(a) 1.71 (b) 1.87 (c) 2.05 (d) 102 (e) 112

6. The SO_2 concentration, in *ppm* by mass, was:
(a) 30 (b) 68 (c) 150 (d) 302 (e) 331

7. The flow rate of SO_2, in actual ft^3/min, at blower exit conditions was:
(a) 0.126 (b) 1.87 (c) 2.01 (d) 4.44 (e) 121

8. If the flow rate of SO_2 is 0.313 lb moles/h, the rate in actual ft^3/h at blower entrance conditions is:
(a) 14,300 (b) 293,000 (c) 653,000 (d) 749,000 (e) 860,000
In another series of tests, SO_2 was added at a rate of 0.290 lb moles/h, and the laboratory reported an analysis of the blower exit sample of 0.006 millimoles SO_2 per liter of sample at standard conditions.

9. The indicated flow rate of air at the blower entrance, in sft^3/min, was:
(a) 28.4 (b) 806.0 (c) 12,900.0 (d) 48,400.0 (e) 774,000.0

10. If the ratio of SO_2 to air is 0.01, and Kp for the $SO_2 \rightarrow SO_3$ is 3.2, the ratio of indicated flow to actual flow is:
(a) 0.392 (b) 0.426 (c) 0.453 (d) 2.21 (e) 2.55

Solution

1. $90 \times (18/14.7) \times (492/520) = 104.3$ sft^3/min. Answer is (d).

2. $0.250 \times (745/760) \times (492/540) \times 0.22331$. gram mols $= (0.22331/22.4) = 0.00994$. Answer is (b).

3. $6.5 \times 0.002 = 0.013$ multiequivalents of SO_2; $(0.013/2) = 0.0065$ millimoles of SO_2; 1 mole SO_2 = 64g; $(0.0065/1000) \times 64 = 4.16 \times 10^{-4}$. Answer is (b).

4. 0.0060 millimoles $\times$ 22.4 ml/millimole = 0.1344 ml of SO_2. (0.1344/1000) = 134.4 ppm by volume. Answer is (b).

5. 20 lb/h $\times$ (1 lb mole/64 lb) $\times$ 359 (sft^3/lb mole) $\times$ (1/60 min) = 1.87 sft^3/min. Answer is (b).

6. Volume % = mole %. Then (150 moles $SO_2/1 \times 10^6$ moles air) is close. $(150 \times 64/ 1 \times 29) = 333$ ppm. Answer is (e).

7. At 80°F and 15 psia, v = $359 \times (540/492) \times (14.7/15.0) = 386$ ft^3/lb mol; 20 lb/h $\times$ (lb mol/64 lb) $\times$ (386 air ft^3/lb mol) $\times$ (1/60 min) = 2.01. Answer is (c).

8. 0.313 moles $SO_2 \times (1 \times 10^6/150$ moles $SO_2)$ = 2084 moles air/h at 78°F and 14.0 lb/in^2 (absolute). v = $359 \times (14.7/13.0) \times (538/492) = 412.2$. Then 2087 $\times$ 412.2 = 860,000. Answer is (e).

9. (0.006 millimoles per liter) = $(0.006/1000) \times 22.4 = 1.344 \times 10^{-4}$; Moles air = $(0.290/1.344 \times 10^{-4}) = 2157$; $2157 \times 359/60 = 12900$ sft^3/min. Answer is (c).

10. $P_{N_2} = 0.76$ atm
$P_{O_2} = 0.24$ atm
$P_{SO_2} = 0.01$ atm

$N_{SO_2} = (0.01/0.24) = 0.0417\ N_{O_2}$
After reaction equilibrium
$N_{SO_2} = 0.0417\ (1 - x)\ N_{O_2}$
$N_{O_2} = (1 - x/2)\ N_{O_2}$
$N_{SO_3} = 0.0417 \times N_{O_2}$

$$P_T = 0.25 \text{ atm};\ N_T = 0.0417\ (1 - x)\ NO_2 + (1 - x/2)\ NO_2 + 0.0417 \times NO_2$$
$$= NO_2\ [0.0417 - 0.0417\ x + 1 - (x/2) + 0.0417\ x]$$
$$= NO_2\ [1.0417 - x/2]$$

$$\frac{0.0417\ x}{0.0417\ (1 - x)\sqrt{\dfrac{(1 - x/2)(0.25)}{1.0417 - x/2}}} = 3.2\ \frac{0.3125\ x}{(1 - x)}$$

$$= \sqrt{\left[\frac{0.25\ (2 - x)}{(2.0834 - x)}\right]\left[\frac{0.09766\ x^2}{(1 - x)^2}\right]} = \left[\frac{0.25\ (2 - x)}{(2.0834 - x)}\right]$$

$$0.3906\ x^2\ (2.0834 - x) = (2 - x)(1 - x)^2; \qquad 0.81383\ x^2 - 0.3906\ x^3 = 2 - 5x + 4x^2 - x^3$$

$$0.6094\ x^3 - 3.1862\ x^2 + 5x - 2 = 0; \qquad x^3 - 5.2284\ x^2 + 8.205\ x - 3.2819 = 0$$

$x = 0.5 \qquad \Sigma = -\ 0.3615$
$x = 0.6 \qquad \Sigma = -\ 0.0251$
$x = 0.6084$
$x = 0.7 \qquad \Sigma = +0.247$

SO_2 remaining $= 1 - 0.6084 = 0.3916$ of ideal. Answer is (a).

THE LICENSING PROCEDURE

It is important for practicing engineers to know that all states and U.S. territories use the same examination; this facilitates registration by comity. The exams are prepared by NCEE. The FE exam tracks the broad-based science, physics, and engineering knowledge that would be gained in a 4-year undergraduate engineering education. A number of questions are repeated on each year's exam to equate its overall difficulty from year to year. If the applicants generally perform equally well with the previous year's examinees on the repeat questions but less well on the total exam, then it is judged to be more difficult than the norm and the passing requirement is adjusted downward slightly. This approach has resulted in an extremely fair test and has avoided some security problems associated with offering identical examinations at each sitting.

The PE exam is a subjective or essay-type exam. That does not mean that the problem solutions do not include calculations, but only that they cannot be machine scored. The composition, scoring plans, and validation of such an exam are challenging.

Composition

What knowledge and skills should an entry-level engineer, with minimum experience, demonstrate in an 8-hour examination? In 1979, NCEE surveyed 4000 licensed professional engineers in practice to identify entry-level knowledge and skills required by each engineering discipline. This is the basis for the examination's content.

Scoring Procedure

A subject expert selects the knowledge and skills to be evaluated by a question and prepares a 10-point item-specific-scoring plan (ISSP). A score of 6 is required for an examinee to be judged to have demonstrated minimum competence in the knowledge and skills being evaluated.

The question statement and its ISSP undergo many reviews before appearing on the exam. All new candidate questions and scoring plans are reviewed by subject experts. The suitable ones are placed in a question box and are the ones from which exams are constructed. A committee of licensed professional engineers, who also may be state board members, is responsible for the final selection of questions. Each exam reflects not only entry-level knowledge and skills, but also a percentage of research, design, consulting, and operation skills as determined by the survey. Completed future exams and ISSPs are reviewed at least twice before being used.

Scoring of the exams is also carefully controlled. Most of it is done by professors from Clemson University, which is near NCEE headquarters; however, a selected group of solutions is also scored by a subcommittee from the Uniform Examinations and Qualifications Commitee (UEQC). Any differences between its interpretation of the ISSP plans and that of the scorer are resolved. This procedure ensures that the scoring adheres to the guidelines of the ISSPs.

An examinee sitting for the PE exam may answer up to eight questions (four in the morning and four in the afternoon). The ISSP for each question allows scores between 0 and 10. The overall passing score is 48, based on a possible raw score of 80.

Validation

NCEE also validates the passing score of 48 for the four major disciplines: chemical, civil/sanitary/structural, electrical, and mechanical engineering. This procedure involves exposing the answers from about 100 examinees to a panel of judges who are practicing licensed professional engineers in the major engineering disciplines listed above.

The panel is asked to answer a single question about each examinee. Has the examinee demonstrated entry-level competence? The results of this judging are then compared with the Clemson scores via a "logistics" analysis which determines what each judge believes is the passing level for the exam.

Thus, the examination parts of the licensing procedure are clearly defined, are well thought through, and have sufficient checks and balances to assure reasonable precision at evaluating an examinee's initial retention of academic knowledge (FE) and entry-level knowledge and skills for minimum competence (PE).

The process for becoming a licensed professional engineer consists of education, experience, examination, and licensure. The rest of the licensing equation—education and experience—is evaluated by each state board of registration according to their individual statutes.

NCEE sends the examinations to each state board for their use. The state boards administer the taking of the examination and return all the examinations to NCEE for grading. NCEE recommends the cutoff or passing score and the state board either accepts that or is free to set what they feel is the minimum score for passing the examination. This flexibility of state boards to change the cutoff score has led to problems with comity but this may be cleared up in the near future. It is very important for the examinee to understand that NCEE does not individually advise the examinee whether she or he has passed or not.

NEW PRINCIPLES AND PRACTICE EXAM IN WORKS AT NCEE

NCEE is developing an objective-response Principles and Practice of Engineering exam, which, if found to be a valid test format and approved by the NCEE Board of Directors in August, could be implemented in April 1988.

The board must decide whether NCEE can prepare properly written test questions that can measure logic and reasoning with the same degree of accuracy as the present free-response testing.

In January, NCEE conducted a "pretest" involving about 700 EITs, to fine-tune the multiple-choice, true-false, and data-selection questions on the exam. For the next two months, experts studied the results of the pretest and improved it to the form of a "prototype." That revised test was given in April 1987 to about 1440 PEs and another 1440 engineering graduates who have taken the Fundamentals of Engineering (FE) exam.

The test consists of questions developed by the NCEE Examination Review Committee, under the supervision of the Educational Testing Service (ETS).

Assuming that the results of the prototype point out that an objective test can replace the essay-type exam presently used, the NCEE board will consider in August 1987 adopting the new format. If approved, beginning with the April 1988 exam, one-fourth of the Principles and Practice of Engineering examination will include objectively scored questions.

The development of an objective-response exam has been an issue at NCEE since as early as 1983. It is believed that the public will benefit by virtue of having a better examination, a more consistent examination, and an examination that will provide an oportunity—by computer—to do a more comprehensive validation study.

Concerning scoring of the current PE test, NCEE believes there is usually more than one way to solve a problem, but that some ways are better than others in terms of economics and protection of the public. The scorer must apply his or her judgment about whether the examinee produced the best possible answer to the problem.

A critique of machine scoreable Principles and Practice of Engineering examinations as being presently contemplated is that:

- Questions do not require necessary refined engineering decisions to obtain acceptable end results.
- Some pendant questions are so simple that they are almost impossible to develop realistically.
- Assumptions made by the problem developer may differ widely from those made by examinees, who can take issue, causing a flood of confusion.
- Small refinements in the correct engineering approach may give different results, but only one answer is correct for scoring purposes.

NCEE believes the cost of implementing a new exam format is undetermined. Much of the expense of the present essay-type exam is accrued during the scoring, which demands individual involvement of engineers who read the answers.

Conversely, an objective-response test is inexpensive to score because results are computer-generated, but it is more costly to develop because it is more difficult to write a good test question that can be scored objectively. In other words, NCEE may be shifting the cost from scoring to development.

SCHEDULE OF FUTURE EXAMINATION DATES

The following administration dates for 1988 through 1999 have been selected as the best choices available in view of the many factors which must be considered in setting examination administration dates. Member boards are reminded that the NCEE staff has been directed to supply examinations only to those member boards administering examinations in accordance with NCEE's examination administration schedule.

Future Dates—Administration of Uniform Examinations

Year	Spring dates	Fall dates
1988	April 14, 15, 16	October 27, 28, 29
1989	April 13, 14, 15	October 26, 27, 28
1990	April 19, 20, 21	October 25, 26, 27
1991	April 11, 12, 13	October 24, 25, 26
1992	April 9, 10, 11	October 29, 30, 31
1993	April 15, 16, 17	October 28, 29, 30
1994	April 14, 15, 16	October 27, 28, 29
1995	April 6, 7, 8	October 26, 27, 28
1996	April 18, 19, 20	October 24, 25, 26
1997	April 17, 18, 19	October 30, 31; Nov. 1
1998	April 2, 3, 4 *or* 23, 24, 25	October 29, 30, 31
1999	April 22, 23, 24	October 28, 29, 30

SOURCE: *1985 Proceedings of the National Council of Engineering Examiners*, courtesy of NCEE.

11

HOW TO PREPARE FOR THE WRITTEN EXAMINATION

Engineering registration examinations are an attempt to discover whether the candidate engineer understands basic scientific and engineering principles and whether his training and experience have taught him how to use these tools to solve problems.

If you have the experience and professional know-how, your registration board *wants* you to have a professional engineer's license. It isn't their job to throw roadblocks in your way. Actually, the opposite is true. We learned before that boards of examiners want to advance recognition of engineering by awarding licenses to the greatest possible number of qualified engineers.

There is little doubt that, as time goes on and the backlog of older unregistered engineers becomes smaller, the number of professional engineers registered by means other than examination will become smaller. It also appears inevitable that it will soon be impossible in most states to process applications except on the basis of written examinations, merely because of the increasing number of applicants.

A board's prime tool for determining whether or not a candidate has the necessary professional know-how is in the written examination. This is a lot to demand of an examination, and NCEE is developing many different types of tests in its efforts to make the exams thorough and fair.

It is the prospect of this written examination which often causes applicants considerable anguish. This is understandable, since the exam is difficult, but it doesn't have to present an insurmountable problem to the average individual.

Knowing what to expect and how to prepare for it will not only ease the fear and mental strain, but it will help the applicant to do a better job in taking the examination.

UNLIKE SCHOOL TESTS

Registration examinations are rarely the kind you remember from college—which aimed at finding out how you had mastered a subject you had just been taught. Registration exams seek to find out knowledge you understand thoroughly and can use—knowledge derived from an entire curricula, rather than knowledge of a single subject.

As we know, the written examination is split into two parts. The first probes into the applicant's grasp of engineering fundamentals of all branches; the second tries to measure ability to apply these fundamentals, as well as to rate the extent and quality of experience in a particular branch or specialty.

All state boards conform on the pattern of the first day's questions. We saw in the previous chapter that these cover that material with which all engineers can be expected to be familiar and which forms the common core of all engineering curricula in college.

EXPERIENCE ON TRIAL IN LAST PART

The problems on the PE exam usually require the engineering candidate to show proper judgment in using correct formulas and practical approaches. Problems are so worded that the candidate must rely on experience to develop a solution. Unlike questions in the first part, which test skill with mathematics and knowledge of engineering theory, the candidate may find an essay-type question or a critical description of a project on which he or she has had experience, putting his or her judgment on trial. Here's an example:

> Give at least four different methods of removing dust particles from a gas and compare their engineering requirements.
>
> Given a gas of 50,000 cu ft per min flow, 50 lb per min of hygroscopic dust, averaging 10 microns in size, at 200°F and with a dewpoint of 50°F, what method would you recommend and why?

An inexperienced person would probably fumble such a problem, while the engineer with only a few years' experience could probably handle it with ease. Since the examination tries to measure the minimum qualifications for registration, questions are usually of this type.

PREPARING FOR THE EXAMINATION

The typical licensing candidate spends a lot of time and money preparing for the examination. Generally, some serious thought in planning this preparation can keep it from being wasted motion. Success in the examination depends in great measure on preparation.

Experience with candidates shows that the average person who fails to pass has tried a "cookbook" approach to the test. She or he will study sample problems, hoping against hope to spot similar ones in the examination itself. Then she or he will only have to insert the variables in a formula and grind out the answer. You will find candidates who spend time studying compilations of sample examinations for frequency and recurrence of certain types of questions, so they can anticipate what particular types will reappear. But it never seems to work out. The cookbook engineer will be lost in the examination.

Although the core of the examination remains essentially unchanged from year to year, the examiners are quite skillful in changing the window dressing, and under the pressure of the test, the cookbook applicant generally misses the basic requirements of the problem and comes up with the wrong recipe.

There isn't any easy substitute for faithful attendance at refresher courses with attention to home assignments in your preparation.

The boards expect candidates to use those sources of information with which they are most familiar. College texts, provided that they are not out of date, are usually good sources of review. The Appendix presents a performance-tested list of books for the various parts of the state examinations for both the Engineer-in-Training and professional examinations. Compilations of past examinations are no longer available from state boards, but typical question pamphlets may be obtained by writing NCEE. The pamphlets are *Fundamentals of Engineering*; *Principles and Practice of Engineering*; *Fundamentals/Principles and Practice of Land Surveying*.

Question pamphlets contain valuable data and acceptable constants which the candidate may use to advantage. A candidate may realize he or she is taking impractical approaches to various types of problems—and in the process of improving his or her approach, the candidate's confidence in his or her own ability will grow.

APPROACHING THE EXAMINATION

There is an art to taking the professional licensing examination, an approach that has worked for many applicants.

All state boards use the open-book examination. Open-book tests measure ability to recognize and apply correctly engineering principles. For open-book tests, the candidate is permitted to use reference books, textbooks, personal notebooks (bound, not looseleaf), and any other material that may be of help. Your state board will advise you about what you are permitted to take into the exam room.

Carry only a minimum number of reference books into the examination room. Some examinees actually carry trunks full of reference volumes to their positions—and then waste most of their time juggling them from knee to knee and thumbing through them, hunting for answers. In addition to losing precious time, they annoy those around them, because examination rooms are generally crowded.

Detailed instructions on what to bring into the examination room are provided by the board. Some means of identification, such as a photograph, is necessary and must usually be presented at the door of the examination room. The following suggested list of materials may be used as a guide:

1. Watch or other timepiece
2. Supply of sharpened pencils and ballpoint pens
3. Battery-operated electronic calculator or computer
4. Rule, scale, triangles, and protractor
5. Reference books as directed

Some of these items are obvious—but that doesn't seem to keep them from being forgotten when they are needed most.

When you get into the examination room, seat yourself in a well-lighted section, if no seat assignments have been allocated. Such care will help ward off fatigue.

TAKING THE EXAMINATION

Before attempting to solve any problems, read them all through and check those which seem most familiar. This should take no longer than 10 minutes. Usually, the candidate's name is not permitted to appear on the exam paper, but a number is allotted each candidate at the start. This should be written in ink on every one of the exam sheets *before* the candidate starts the examination.

After reading over the questions, divide your remaining time equally among the problems which you decide to work out, and adhere strictly to these time allotments. The instructions on the examination sheet will tell how many problems must be solved successfully. NCEE rarely chooses problems that take more than 20 to 45 minutes to solve, and many take less. Use this cue and do not go beyond the time limit. A sand timer calibrated to a half-hour period has been found useful. Some people have taken clock timers and even alarm clocks into the examination room, but this practice is to be discouraged for obvious reasons of disturbance.

In most states, answers receive equal weight, regardless of the time spent on them by the examinee. Then why spend more than the allotted time? If a problem takes more, drop it and go on to the next. Of course, if you are near the end and could finish it in a few minutes, do so and then go on. But be sure not to become ensnarled for longer than 45 minutes in any one solution. Others will take less time, and it will be possible for you to go back. Some boards are including hints and suggestions directly on the examination paper to help and guide the candidates to successful completion of the paper.

Work problems that seem simplest first, and do one solution or a part of a solution on a single sheet of work paper. This helps you save time if you should skip around. See Chapter 10.

Solve or attempt to solve the full requirement of problems. The boards look to the candidate to reflect breadth and scope by such a practice. If you cannot finish a solution, give as much of the solution as possible in the time allotment. So long as the *method of solution* is reasonable, you will get credit for the part worked. If time is running out, set down the method of solution step by step and carry through the numerical solution as far as you can go. To save time, when the final answer is obtained by means of solving a complicated formula, merely fill in the terms with the numerical values and place the dimension of the answer at the end. Then go on to the next problem. If there is time, go back and work out the numerical answer.

A common complaint in examination "post-mortems" is that the engineer just could not recall the simplest formula or the most familiar constant. The candidate was well prepared, but just could not respond. He or she had run into an "emotional block."

This is a common occurrence, so it is best not to worry about it. Just take a few common-sense precautions and relax.

1. Don't try to finish your preparations the day before the test or stay up late trying to cram.
2. Take a mild form of exercise the day before.
3. Get to bed early and have a full 8- to 9-hour sleep.
4. Get to the examination a half hour ahead of time.

During the examination, should tension mount, take a break for a minute or two. If permissible, leave your seat and walk around, to relieve tension. When you return, reread the problem before you begin to work it again, or try another first.

But careful preparation that builds confidence in your ability to pass the examination is probably your best defense against the emotional block. A good understanding of fundamentals will not trap you into a hollow sense of security.

Examination Checklist

1. Read directions carefully.
2. Read problems first, before attempting to answer.
3. Allot time to each problem and keep close check.
4. Do problems with which you are most familiar first.
5. Show sketch or diagram with each solution.
6. Strive for the full complement of problems solved.
7. Method of solution is given most weight.
8. Strive for a reasonable final answer.
9. Reread problem after solution is completed.
10. Outlining the solution step by step is given credit to a degree.

Write to NCEE for free copies of their *FE Memorandum* dated November 10, 1985, and NCEE's *Principles and Practice Examination Specifications*, effective with the fall 1986 administration and revised April 1986.

WHERE TO WRITE FOR INFORMATION*

- National Council of Engineering Examiners, Box 1686, Clemson, SC 29633-1686 PE Study Guides: examination books, typical questions pamphlets, sample examinations booklets.
- National Society of Professional Engineers, Information Center, 1420 King Street, Alexandria, VA 22314. *Selected Bibliography on Professional Engineers and Fundamentals (Engineers-in-Training) Examination Preparation Including Study Manuals and Home Study Courses.* NSPE Publication No. 2201, latest revision.

---

* Courtesy of NCEE.

Accredited Engineering Programs

Accreditation Board for Engineering and Technology (ABET), 345 East 47th Street, New York City, NY 10017. Request list of accredited engineering programs and engineering schools where accredited engineering programs are offered.

Videotaped PE Refresher Courses

- Association for Media-Based Continuing Education for Engineers, Inc. (AMCEE), 225 North Avenue, NW, Atlanta, GA 30332.
- College of Engineering, Iowa State University, Continuing Education, 240 Engineering Annex, Ames, IA 50010.
- Garland H. Duncan, PE, Chairman, Region XI ASME, Professional Development, P. O. Box 3091, Tequesta, FL 33458.
- U.S. Air Force Academy (USAFA). Contact HQ USAFA/DFCE, U.S. Air Force Academy, Colorado Springs, CO 80840-5841. Defense Audiovisual Agency (DAVA). Videotaped course Engineering EIT Refresher Course: Air Force audiovisual numbers VT34C602835-VT34C602864.

STUDY HELPS AVAILABLE FROM NCEE

Examination Books—Volumes I and II

Vol. I, *Professional Engineering Examinations 1965–1971*

Vol. II, *Solutions to Professional Engineering Examinations 1965–1971*

Study Guides

Professional Engineering Examinations 1972–1976

Fundamentals of Engineering (FE) Sample Examination 1983

Principles and Practice of Engineering (PE) Sample Examinations Group 1, 1984

Typical Exact Questions (pamphlets)

Relating to the Fundamentals of Engineering Examination

Relating to the Principles and Practice of Engineering Examination

Other Publications

The Registration of Professional Engineers and Land Surveyors in the United States

Model Law

Model Rules of Professional Conduct

Proceedings (annual)

Registration Bulletin

12

WHAT TO LOOK FOR IN REFRESHER COURSES

You have filed your application and have received word from your board that you have qualified to sit for the written examination. Now you want to take a refresher course to prime yourself for the examination.

Write to your state board secretary for the names of technical and professional organizations offering such courses in your area. If refresher courses are inaccessible or inconvenient for some reason, there are home-study refresher courses and home-study aids available. Again, your board secretary is in the best position to advise. Contact the local chapters of your state professional engineering societies. Many courses are springing up throughout the country and information is available.*

What do you look for in a refresher course? Remember, they are fast-moving and require attentive and religious attendance and ability to get homework assignments done. Most such courses can only show what material is to be stressed in more thorough home study. Do not expect merely to sit in and come away with all the know-how necessary to work out examination problems.

PREREQUISITES

The organization of refresher courses assumes that students or candidates have a working knowledge of the subject matter to be covered. It is expected that they have a working knowledge of simple mathematics

---

* Constance, John D.: "A Guide to P.E. Refresher Courses," *New Engineer*, October 1977.

and the physical sciences and are employed in engineering work and not in positions merely associated with engineering. Since the examination stresses design, they are expected to be familiar with the principles of design. Above all, they must be willing to work at the course and not to procrastinate, expecting to catch up in the last month. Then, it will be too late.

Student candidates should have met the minimum requirements set up by their board of examiners as to age, education, citizenship, and experience. They should also have some knowledge of the general licensing picture in their state. Review the chapters on requirements, qualifying experience, evaluation criteria, and filing procedure. We will repeat the checklist appearing at the end of Chapter 6. Go over it with the course adviser during the interview prior to registration for the course.

1. Are you over twenty-five (minimum age of nineteen in New York)?
2. Are you a citizen of the United States?
3. Are you a high school graduate or equivalent?
4. Are you presently employed in engineering work?
5. Do you have a degree in engineering?
6. How many years of experience have you in engineering work?
7. Are you in responsible charge of work or people?
8. Are you personally acquainted with at least three PEs?
9. Do your responsibilities consist solely of repetitive operations?

Also use the self-evaluating checklist under "Are You Ready for Professional Registration" in Chapter 6.

COURSE ORGANIZATION

Look wherever possible for a course approved or registered by your state board of examiners. A course adviser should interview each potential registrant to learn if she or he has the necessary prerequisites. Look for completeness here.

Look for individual instructors for each subject. The instructors should be registered professional engineers and experts in their respective fields. These are the earmarks of good and well-established organization. Some courses select their instructors from industry, others from the faculty of nearby colleges and universities.

Complete descriptive instructions and course outlines are a necessity for students to follow. Such instructions may cover attendance requirements, prerequisites, course coverage, books, note-taking, schedule of lectures, application deadlines, and examination dates.

Crowded classrooms should be avoided. Poor lighting and lack of adequate blackboard space are not conducive to an attentive student audience. Look for these telltale signs. Request permission to sit in on a lecture or two before signing up. Refusal of such a request may indicate unsatisfactory conditions.

REFERENCE BOOKS

Most refresher courses tend to become cram sessions, accompanied by feverish note-taking. To lessen the strain, study aids especially written for course coverage are often available in many areas. If not, contact the local bookstore or the local chapter of the state society of professional engineers. Compilations of typical examinations are a necessary tool in refresher courses, to check the past and to anticipate the trend for the future. A number of recognized standard reference textbooks and study aids are listed in the Appendix. Whatever the student uses, he or she should be very familiar with them.

ENGINEERING FUNDAMENTALS

Some persons will require pre-refresher-course preparation. To this end a description of a typical refresher course follows. In a refresher course, the basic principles underlying the subject of hydraulics, machine design and mechanics, thermodynamics, and electricity are fully explained, discussed, and illustrated by the solution of problems.

Topics in *hydraulics* cover pressure in a fluid, pressure gauges, Archimedes' principle, stability of a ship, the hydrostatic paradox, forces against a dam, properties of a surface, coefficient of surface tension, angle of contact, capillary rise in tubes, formation of drops, streamline flow, Bernoulli's theorem, discharge rate of a pipe, and flow of viscous fluids through tubes.

The topics in *mechanics and machine design* cover composition and resolution of vectors, moments, center of gravity, uniform motion, Newton's second law, path of a projectile, work and energy, impulse and momentum, rotational motion, moment of inertia, translation and rotation, elasticity, harmonic motion, and gravitation.

The section on *thermodynamics* covers temperature, thermometers, linear expansion, thermal stresses, heat, the mechanical equivalent of heat, specific heat, calorimetry, heat of combustion, internal energy,

Carnot cycle, Otto cycle, Diesel cycle, heat transfer, properties of gases, entropy, enthalpy, and application and use of steam tables.

The topics in *electricity* cover Coulomb's law, the electric field, electric potential, current, resistance, resistivity, direct currents, magnetic field, galvanometers, ammeters, voltmeters, the direct-current motor, inductance, ferromagnetism, alternating currents, basic electronics.

The normal performance-tested total hours of instruction is 48 hours, or sixteen 3-hour sessions.

BASIC MATHEMATICS AND PHYSICS

A course in this area is a review of basic mathematics and physics required of engineers, with direct application to the problems encountered by the practicing engineer.

The topics in *mathematics* should include: elementary algebra, geometry (plane), trigonometry, solid geometry, analytical geometry, differential and integral calculus.

In the subject of *physics*, the topics should include: elementary and engineering mechanics, hydraulics, introduction to mechanics of materials, electricity and magnetism, elementary ac and dc circuit theory, motors, power transmission, and basic electronics. Thermodynamics includes concept of temperature, work, ideal gas law, Carnot's theorem, Carnot cycle, Otto cycle, entropy, enthalpy, Mollier diagrams, and steam tables.

Acceptable course-presentation coverage is 48 hours, or sixteen 3-hour sessions. A number of states do not review mathematics and physics, as such.

CHEMISTRY

Review of this subject would include general chemistry, industrial stoichiometry, and water treatment. Total hour coverage would be 30 hours, or ten 3-hour sessions. Use of past examination problems will provide material for presentation. Some states do not offer chemistry problems.

BASIC ENGINEERING SCIENCES

So far, we have outlined course coverage equivalent to preparation for the first 4-hour examination on the morning of the first day. Refer to Table 10.1 for complete coverage of the FE examination.

BRANCH OR SPECIALTY

For the PE examination, see Table 10.2. You should also write to NCEE for example examination pamphlets for disciplines of interest.

HOW TO GET READY FOR REGISTRATION

If the prospect of a difficult, lengthy examination has kept you from becoming licensed, don't despair. If you meet the statutory requirements for registration, you have probably retained more than you think of the engineering fundamentals you learned in school or picked up through experience. If you've been out of school for many years, or you have specialized rather narrowly, or you have been an administrator too long, but are determined to become licensed, help is available for getting past the obstacle of the state examinations.

First, there are specially tailored refresher courses. These are given by colleges and universities, state chapters of the National Society of Professional Engineers (NSPE), and technical societies. Write to NSPE for their Publication 2218, *State-by-State Guide to Fundamentals of Engineering and Principles and Practice of Engineering Review/Refresher Courses.*

If a refresher course is not convenient, you might consider a correspondence course. Again, write NSPE for their Publication 2201.

Finally, you might consider reading on your own, for there are quite a number of helpful books written by engineers who have passed the licensing exam and have engaged in helping other engineers pass it. NSPE, again, has compiled a bibliography of such books, NSPE Publication 2201. As a matter of fact, the author suggests you write NSPE for their *Subject Listing of NSPE Publications*, revised July 1986. Also write NCEE for a listing of study guides available.

Before you sign up to take a refresher course or purchase any study helps, you should make certain you qualify to sit for the FE or PE examinations. And before you sign on the dotted line and pay, you should evaluate the course. One way of doing this is to compare its coverage against the example questions available from NCEE (also look for help from the books listed by NSPE in Publication 2201).

It should be emphasized that most courses are truly refresher courses. A great deal of knowledge of the subject matter is expected. The pace is usually fast, and students are expected to keep up and do their homework.

Before enrolling, check the course coverage, the qualifications of the instructors (if not registered engineers, they should be recognized experts in the discipline they will instruct), the prerequisites (worthwhile courses require a degree in engineering or a number of years of engineering experience), and whether the course is recommended by the state NSPE organization or another state-approving agency.

And remember, merely sitting for a course and not doing homework problems will be a waste of time and money.

13

MULTIPLE REGISTRATION

Engineers simply cannot accept state or regional boundaries as limitations on their activities, because of the nomadic nature of engineering practice itself, but a license to practice professional engineering cannot be transferred from one state to another: there is no such thing as national registration for engineers. An applicant may secure registration in another state by *endorsement*, but the registration is not transferred. Each state retains the right and considers it its duty to regulate engineering within its borders for the protection and welfare of its citizens. To do so, it must pass upon the qualifications of all engineers who seek to practice there—whether licensed elsewhere or not. While, on the one hand, the various state boards are charged with promoting and facilitating the flow of engineering talent across state lines, on the other, they must prevent the use of interstate registration as a means of getting around the different licensing requirements of any other state.

Thus, the engineer or consultant is frequently confronted with the necessity of obtaining registration when she or he accepts an assignment in a state in which she or he has not previously registered. This becomes a perplexing problem and can cause undue hardship.

TEMPORARY PRACTICE

Most states permit a registrant of another state to practice temporarily, if the engineer has only an occasional assignment. The limit of temporary practice, without permanent registration, varies among the states. Usually, the engineer can work from 15 to 60 days in any one calendar year without applying for permanent registration. In some cases, state boards will permit temporary practice beyond the legal nominal limit—pending the processing of an application for permanent registration.

Interested applicants should write boards for particulars regarding this aspect of licensure.

These provisions appear in six categories in the various registration laws as follows:

1. Nonresident—not exceeding 30 days in any one calendar year
2. Nonresident—not exceeding 60 days in any one calendar year
3. Nonresident or recent arrival in state—more than 30 (or 60) days in any one calendar year if application for registration has been filed, such exemption to continue only during time required for consideration of application
4. Nonresident or recent arrival in state—during such time as may be required for consideration of application
5. Temporary permit without project limitation (except as indicated)
6. Special permit limited to a specific project and renewable annually until project is completed

The provisions of some registration laws pertaining to temporary or special permits are conflicting, somewhat confused, and, in many instances, burdened with arbitrary, capricious, and discriminatory classifications which endanger the constitutionality of the registration laws in which they appear.

Most agree that a provision in a registration law which grants the privilege of practice without registration during pendency of application to a nonresident or to a recently arrived resident and denies that privilege to resident applicants is, beyond a doubt, unconstitutional. This is supported by the opinion of the Supreme Court of Pennsylvania in *Commonwealth v. Humphrey*, 288 Pa. 280, holding the first registration law of that state unconstitutional on the sole ground that it exempted from registration engineers employed by corporations engaged in interstate commerce and required registration of all engineers not so employed.

LICENSURE BY ENDORSEMENT

To promote and facilitate the free flow of engineering talent across state lines, most states provide for *licensure by endorsement* for those engineers previously registered in one or more other states. Although the requirements differ to some degree with each state, written and oral examinations are waived by most boards if the applicant has passed all parts of a reasonably equivalent written examination in the state of prior registration. Most states require that applicants be registered in their home state. Applicants must, of course, also meet the statutory requirements of age, citizenship, and experience.

Before a state board will approve an applicant for a license under the provision of licensure by endorsement, it will always try to make certain that the applicant is fully qualified. Specifically, applicants may be considered for endorsement if they hold 1) a license or certificate to practice professional engineering issued after examination by a legally constituted board in any other state or political subdivision of the United States, or 2) a Certificate of Verification issued by the National Council of Engineering Examiners (NCEE), *provided* the following are true: (1) that when the license or certificate was issued, the issuing state's examination was the full equivalent of the examination in the state in which the application is to be filed, at the time of prior registration; and (2) that the applicant's record fully met all other statutory requirements—age, citizenship, education, experience, and character.

When a board receives an application for registration by endorsement, it is individually examined for verification of record to see that board standards have been met as of time of prior registration. The applicant must show documentary proof that all legal requirements have been met before a license is granted. To the extent of her or his record and accomplishments, the engineer-in-training is accorded the same treatment as the registered engineer.

MULTIPLE REGISTRATION BY CERTIFICATE OF VERIFICATION

It wasn't too long ago that an engineer had to appear personally before state boards of registration even though the engineer's qualifications were beyond question. Fortunately, this has mostly been eliminated, due in large measure to the work of the NCEE. Its Committee on Records Verification has endeavored to eliminate from state registration laws the requirements having little purpose other than to add to the difficulty of obtaining a license. NCEE continues to foster reciprocity and licensure by endorsement among state registration boards.

The Records Verification Program serves as an agency for engineers needing registration in more than one state (multiple registration), to minimize the effort, avoid the embarrassment, and reduce the expense required of registered engineers when obtaining licensure in other states.

Licensed engineers in this situation will find their applications greatly expedited and, in many cases, promptly acted upon when they have a Council Record. The bylaws of NCEE and the registration laws of most states specifically provide for recognition and acceptance of the Certificate of Verification as evidence of qualifications for registration as a professional engineer.

Procedure and Requirements

Registered engineers desiring a Certificate of Verification must submit to NCEE on a prescribed form their complete record, including education, engineering experience, legal registration, professional examinations, membership in technical societies or professional societies, and references for character and experience. All information and application forms may be obtained from:

National Council of Engineering
Examiners
P. O. Box 1686
Clemson, SC 29633-1686

Upon receipt of your application forms, NCEE will review them and initiate their processing. The subsequently compiled Council Record contains a concise report of the engineer's education, experience, written examination results, PE references, and registration which are verified and maintained by NCEE. The original is kept at NCEE headquarters in Clemson, South Carolina. A copy of your Council Record will be transmitted to other registration boards only with your written permission; records will not be released to any other agency or person.

Each engineer maintaining a Council Record will receive a free subscription to the *Registration Bulletin*. This newsletter is published five times a year and contains pertinent and timely information of interest to professional engineers.

The Records Verification Program (RVP) is a fact-finding and verifying agency which serves as a clearinghouse for the professional records of licensed engineers. It reviews an engineer's record of experience and investigates licensure status with the home state board. Once all requirements are met, NCEE will issue a Certificate of Verification.

The certificate serves as a verified record of an engineer's experience. It helps competent engineers to cross state lines and eliminates duplication of effort by state boards. Because the compilation of information takes time, an engineer should not wait until it is needed (usually in a hurry).

The certificate must be updated annually to remain in force.

RECORDS VERIFICATION PROGRAM AND QUALIFYING EXPERIENCE

The Certificate of Verification will be issued only to applicants whose professional records contain conclusive evidence that they are fully qualified and competent to practice professional engineering and that they have actually had responsible charge of engineering work of magnitude and complexity and have made important engineering decisions therein.

The basic objective of requirements of qualifying experience is to ensure that the applicant has acquired, through actual practice of suitable caliber in engineering, the professional judgment, capacity, and competence in the application of the engineering sciences requisite to registration.

Quality. Experience shall be of such quality as to demonstrate that the applicant has developed technical skill and initiative in the correct application of engineering science, sound engineering judgment in the creative application of engineering principles and in the review of such applications by others, and capacity to assume responsibility for engineering work of professional character.

Scope. Experience shall be of sufficient breadth and scope to ensure that the applicant has attained reasonably well-rounded professional competence in his or her basic engineering field, rather than or as well as highly specialized technical skill in a very narrow and limited branch of that field.

Progression. The record of experience should indicate successive and continued progress from initial work of simpler character to recent work of greater complexity and higher degree of responsibility, and continued interest and effort on the part of the applicant toward further professional development and advancement.

Capacity. The record of experience should indicate that the applicant has attained to a considerable degree those attributes of clear thinking and keen analysis essential to professional competence. It should demonstrate a capacity for orderly assimilation, evaluation, utilization, and resolution of engineering data; and an ability to reach sound decisions, obtain safe and economical results, and render reliable and authoritative professional engineering advice (to client or to employer).

The determination of the applicant's competency will be based less on length of experience (provided the minimum is met in each case) than on the breadth, quality, and importance of pertinent experience, degree of progression, evidence shown of ability to accept increasing responsibilities, and the possession of personal qualifications required for successful engineering work. Responsible engineering teaching may be construed as engineering experience.

The term "practice of engineering" shall mean any professional service or creative work requiring engineering education, training, and experience, and the application of special knowledge of the mathematical, physical, and engineering sciences. This includes work in consultation, investigation, evaluation, planning, design, and supervision of construction for the purpose of assuring compliance with specifications and design, in connection with any public or private utilities, structures, buildings, machines, equipment, processes, works, or projects.

RVP interprets that the practice of engineering does not include the work ordinarily performed by persons who operate or maintain machinery or equipment, or whose only experience has been in the execution, as a contractor, of work designed by a professional engineer or the supervision of the construction of such work as a foreman or superintendent.

ENGINEERING PRACTICE IN ANOTHER STATE—PROBLEMS AND PITFALLS

There have been a significant number of cases in the various states dealing with individual engineers, or engineers practicing in the partnership form, where the client sought to avoid paying the engineer on the grounds that the design contract was "illegal," since the engineer was unlawfully practicing in the jurisdiction. In general, the engineer has been denied fees under such circumstances except in cases where the engineer was working for another design professional who was qualified to practice in the jurisdiction in which the project was located. The distinction in these cases involving individual engineers seems to be that the licensing statutes were enacted for the protection of the public and not for the protection of a fellow design professional.

Licensing of Individuals

The state licensing laws make it clear that it is unlawful for an individual to practice engineering within the state unless that person is duly licensed and registered in accordance with state law. The laws are much less clear as to what constitutes the practice of engineering in the state.

Further complexities arise with respect to statutory provisions which provide for limited practice in State B by an individual who is licensed only in State A. Several states have exceptions in their licensing laws which grant limited permission to practice to nonresident engineers who have no established place of business within the state. For example, New York, New Jersey, North Carolina, and South Carolina will give a limited permit to practice for 30 days to a nonresident, provided that person is legally qualified to practice engineering in his or her home state. Michigan's practice limitation extends to 60 days, with the additional requirement that the nonresident's own state grants reciprocity to Michigan engineers. Another variation is found in West Virginia's statute, which will grant a permit to practice on a specific project for the duration of the project. The limited practice candidate can avoid licensure, but only if the time limit is not exceeded. To continue to practice in State B, the individual thereafter must file for a certificate of licensure or run afoul of the registration laws.

What Is Engineering Practice?

Legally, it is arguable that an individual engineer registered in State A, who does not physically render his or her services in State B (state in which the project is located) and who designs (while residing in State A) a project to be built in State B, is not practicing engineering in State B. Without some other provision in State B's statutes (e.g., requiring the sealing of plans or precluding public agencies from contracting to have work performed by nonlicensed—in State B—engineers), State B has no jurisdiction over the State A engineer. In any event, most states do have provisions for a foreign engineer, not licensed in that state, to render limited services within the state.

Despite State B's apparent lack of jurisdiction over the State A engineer in the foregoing, a State B client nevertheless may seek to avoid paying the engineer on the ground that the engineer was illegally performing services in State B. The nonpaying client often has been successful in such cases, which as a general rule must be tried in State B.

Affixing of Seal Requirement

Various state and local statutes require that plans for certain construction bear the seal of a professional engineer licensed in that state. Such requirements are straightforward and usually present no insurmountable problems; however, it should be noted that a State A engineer who designs a project in State B and arranges for a State B–licensed engineer to review and seal the plans may still be in violation of the licensing statute.

Government-as-Client Limitation

The licensing statutes include provisions, other than the requirements of licensure to practice in the state and the sealing of plans, which have a direct bearing on the performance of work in connection with an out-of-state project. Most licensure statutes require that governmental entities have their engineering services performed by engineers licensed in the state. Some states require only that the plans be prepared by the in-state licensed engineer; others prohibit the agency from contracting with non-in-state licensed engineers. But here arises a difference. Under the first restriction, the State A engineer apparently could contract for work and then (if there is no contractual prohibition against subcontracting) subcontract to a licensed State B engineer. But such procedures must be carefully watched for infringement against state statutes.

Corporation Limitations

Some states permit domestic business corporations (i.e., other than professional corporations) to practice professional engineering, provided that the engineering work is performed under the direct control of licensed professional engineers, and often requiring that certain of the officers, directors, etc., of the business corporation be licensed. As a general rule, an out-of-state business corporation will be permitted to practice engineering in State B to the same extent as a domestic State B business corporation, provided that the out-of-state business corporation is also permitted to practice under its home-state laws.

Virtually all states* permit professional corporations to practice engineering. Generally, all officers, directors, and stockholders of a professional corporation must be licensed professional engineers.

With regard to practice by an out-of-state professional corporation, most of the licensing statutes are either silent or ambiguous. The laws relating to professional corporations have been enacted in most states but need to be subjected to a greater amount of interpretation. Even

* Not New York State.

when these relatively new acts are read in conjunction with the business corporation and the licensing statutes, some uncertainty arises in determining if a foreign professional corporation can practice lawfully within a particular state.

Separate from the question about the practice of engineering in a foreign state is the question of whether a corporation (whether business or professional) formed in State A may conduct any business in State B. This question usually will arise even though all of the stockholders, directors, and officers of the State A corporation are individually licensed in State B. A corporation is an artificially instituted being created by the state of its incorporation and has no legal existence outside of the state of corporation (unlike an individual, who legally exists everywhere).

The need for the State A corporation to qualify to do business in State B is independent of the need for the State A corporation to qualify under the professional corporation laws. Thus, for example, in the states which permit corporate practice of engineering as long as the persons in charge of the work are licensed, a State A corporation can readily comply with the licensing statute (if the proper individuals are licensed in State B). However, such qualification under the licensing laws does not mean that the State A corporation can lawfully conduct business in State B.

In the case of a business corporation, all states make provision in their corporate laws for the authorization for a foreign corporation to do business in the state. The principal limitation that State B places on the State A corporation is that it may conduct business (such as it may be) only as is permitted a State B corporation. Other requirements to qualify in the foreign state are largely formalistic, i.e., the filing of certain information, payment of taxes, etc.

Summary

In most jurisdictions there are three sets of relevant statutes: the licensing statutes, the business corporation statutes, and the professional corporation statutes. These statutes are administered by at least two separate agencies of the state and the statutes are generally uncoordinated with each other. Accordingly, as a result, there are many unresolved questions with respect to the practice of engineering in a foreign state by a professional corporation.

It is suggested that the interested professional engineer write to NSPE for a copy of their Bulletin 2217 entitled *State-by-State Summary of the Engineering Corporate Practice Laws* as well as their subject listing of publications.

14

YOUR PE ENGINEER'S SEAL

All states and territorial possessions authorize or require professional engineers to have a personal seal or stamp, and prescribe the requirements and limitations of its use.

In some respects, the use of the seal by a properly registered professional engineer seems to be very little understood. State registration laws declare that licensed engineers may have a seal and that plans, specifications, and reports prepared by or under the supervision of such licensed individuals shall be stamped over the personal signature of the registrant when filed with public officials. A few states require the registrant to obtain a seal prior to issuing the certificate of registration.

The professional seal serves a double purpose; it identifies the practitioner whose name and number it bears, and it indicates that that person has obtained state authorization to engage in professional practice. In the first instance, it establishes responsibility for the work with which it is connected, and, in the second, it stands as a hallmark of an acceptable degree of professional competence.

State registration laws, in protecting public property and life, may revoke a professional engineer's license if he or she permits the seal to be affixed to any plans or specifications that were not prepared by him or her or under that person's personal supervision by regularly employed subordinates. Often, licensed engineers are asked to examine drawings made by others, not their regularly employed subordinates. Unless they satisfy themselves as to the engineering correctness of the design and drawings by going over them in detail before affixing their seal, they render themselves liable; and should injury or death occur as a direct result, they may be brought up on charges of criminal negligence. The person who prepared the plans, being unlicensed, may be brought up on charges, fined, and/or imprisoned.

The time may be not too far off when one of the effects of Engineers' Registration will be manifested in a professional engineer's use of her or his seal. Industrial concerns are beginning to make it a requirement for their professional engineers with licenses to affix their seals to a design drawing developed by them personally. This policy not only develops the professional consciousness of the engineer but also protects the employer against charges of negligence should failure occur and life be lost or property damaged.

As firms become larger, those in responsible charge of engineering, as department heads or chief engineers, etc., are rarely able to supervise personally the work of their subordinates. As a result, they are becoming more insistent that qualified subordinates become licensed, so they can carry on the work more efficiently. Work schedules should be arranged so there is time for the examination of plans and specifications by a duly licensed engineer in their employ.

Licensed engineers are cautioned against the practice of affixing their seals to drawings not in their specialty. Persons licensed as professional engineers can easily fall prey to such practice. Their specialty may be structural or civil engineering, but they may be called upon to affix their seal to a mechanical or piping drawing. Although this may be done on occasion in larger organizations as a last-minute expedient, it should be watched and avoided under penalty of the law.

Should a firm which has been lawfully practicing professional engineering in the state of its incorporation retain an engineer in another state to affix his or her seal to drawings made by that firm in order that the plans may be used in the engineer's state, both the engineer and firm become liable. The engineer may lose his or her license for aiding and abetting unlawful practice and for affixing a seal to drawings which he or she did not prepare. The firm may be taken into court for unlawful practice.

SEAL IS PERSONAL PROPERTY

The seal is a very personal thing. It is the personal property of the engineer and not the property of those for whom she or he works. The seal is equivalent to the engineer's signature and determines responsibility to the public. Laws authorizing or requiring the use of a seal usually require its use in connection with the signature of the person to whom it belongs. And in view of the ease with which duplicate seals may be made up, licensed engineers would be serving their own interests, protecting their own names, and maintaining the integrity of their registration, if, before affixing their seal to a print of a drawing executed by them, they would sign their name just as it appears on their registration certificate. Then they should stamp directly adjacent to the signature.

The seal may not be given, loaned, or sold to any other person for professional purposes, nor may it be acquired, appropriated, or used by another. Its use is mandatory on plans to be filed with public officials, and no public official may lawfully accept any plan or specification to which a seal has not been affixed.

Any question as to the use of the seal shall be referred to the secretary of the board of examiners. That person alone has the right to give official information on the use of the seal.

And, remember, "your seal is a very personal thing which is reserved for your work or work prepared under your direct supervision by your regularly employed subordinates. It cannot lawfully be used elsewhere." Guard its use with care. Why jeopardize your hard-earned responsibility?

MODEL LAW REVISIONS ON USE OF ENGINEER'S SEAL

From information received from state boards, the Model Law Revision Committee of NCEE found it obvious that no uniformity existed among the various states. Some states did not require use of the seal; others required it to be used in various ways for various engineering documents. Some required a signature, while others did not. After formulating a comprehensive list of state policies, the committee worked out a consensus position, embodied in the following recommendations:

1. The seal shall be a rubber stamp.* Whenever the seal is applied, the registrant's written signature shall be affixed adjacent to the seal. No further words or wording are required. A facsimile signature will not be acceptable.
2. The seal and signature shall be used on all drawings, plans, design information, and calculations whenever presented to a client or any public or governmental agency. Specifications and reports shall also be included.
3. The seal and signature shall be placed on all original copy, tracings, or other reproducible documents in such a manner that the seal and signature will be reproduced. The application of the registrant's seal and his signature shall constitute certification that the work thereon was

* New Jersey uses an embossing seal; it is illegal to use rubber stamps.

done by him or under his control. In the case of multiple sealings, the first or title page shall be sealed and signed by all involved. In addition, each sheet shall be sealed and signed by the registrant or registrants responsible for each sheet. In case of a firm, partnership or corporation, each sheet shall be sealed and signed by the registrant or registrants involved. The principal in responsible charge shall sign and seal the title or first sheet.

4. The seal and signature shall be used by registrants only when the work being stamped was under the registrant's complete direction and control.

5. In the case of a temporary permit issued to a registrant of another state, the registrant shall use his state of registration seal and shall affix his signature and temporary permit to his work.

6. In the case of a registrant checking the work of an out-of-state registrant, the state registrant of the use state shall completely check and have complete dominion and control of the design. Such complete dominion and control must include possession of the seal and the signed reproducible construction drawings, with complete signed and sealed design calculations indicating all changes in design.

7. The design of the seal shall be determined by each state board. However, the following minimum information shall be on the seal:

State of registration
Registrant's name
Registrant's license number
The words "Licensed Professional Engineer"

In committee discussion several objections were raised. The chief objection was to the signing of the original documents where one might have to turn them over to a client or a governmental agency. The question came up concerning the possibility that, under those conditions, the original tracings might be altered to the detriment of the responsible engineer who performed the original work. It was recommended that the engineer who prepares the work should keep a copy of all work of reference. Then, should a question arise as to alteration of the work, the engineer could always pull out his or her copy.

CERTIFICATION OF DOCUMENTS VERSUS PLAN STAMPING

The followng differences between "certification of documents" and "plan stamping" should be pointed out as causes for professional liability suits:

- State board rules require that a licensed design professional certify only those documents prepared by the licensee or prepared under the licensee's direct supervision.
- Plan stamping is an act where a licensee certifies documents prepared by others not under the control of the licensee and where the licensee had little or no involvement in preparing the documents. Plan stamping is in violation of state statutes and violators are subject to disciplinary action.

APPENDIX

Addresses of the National Council of Engineering Examiners and Its Member and Affiliate Member Boards

National Council of Engineering Examiners
P.O. Box 1686, Clemson SC 29633-1686
(803) 654-6824
1985

Alabama

State Board of Registration for Professional Engineers and Land Surveyors
Suite 212, 750 Washington Avenue, Montgomery, AL 36130
Attn. Executive Secretary
Telephone: (205) 261-5568

Alaska

State Board of Registration for Architects, Engineers and Land Surveyors
Mail: Pouch D-LIC
State Office Building, 9th Floor
Juneau, AK 99811
Attn. Licensing Examiner
Telephone: (907) 465-2540

Arizona

State Board of Technical Registration
1645 W. Jefferson Street, Ste, 140
Phoenix, AZ 85007
Attn. Executive Director
Telephone: (602) 255-4053

Arkansas

State Board of Registration for Professional Engineers and Land Surveyors
P. O. Box 2541
Little Rock, AR 72203
Attn. Secretary-Treasurer
Telephone: (501) 371-2517

California

The Board of Registration for Professional Engineers and Land Surveyors
1006 Fourth Street, 6th Floor
Sacramento, CA 95814
Attn. Executive Officer
Telephone: (916) 445-5544

Colorado

State Board of Registration for Professional Engineers and Professional Land Surveyors
600-B State Service Bldg.,
1525 Sherman Street
Denver, CO 80203
Attn. Program Administrator
Telephone: (303) 866-2396

Connecticut

State Board of Examiners for Professional Engineers and Land Surveyors
The State Office Building, Room G-3A
165 Capitol Avenue
Hartford, CT 06106
Attn. Administrator
Telephone: (203) 566-3386

Delaware

Delaware Association of Professional Engineers
2005 Concord Pike
Wilmington, DE 19803
Attn. Executive Secretary
Telephone: (302) 656-7311

District of Columbia

Board of Registration for Professional Engineers
614 H Street, N.W., Room 910
Washington, DC 20001
Attn. Executive Secretary
Telephone: (202) 727-7454

Florida

Board of Professional Engineers
130 North Monroe Street
Tallahassee, FL 32301
Attn. Executive Director
Telephone: (904) 488-9912

Georgia

State Board of Registration for Professional Engineers and Land Surveyors
166 Pryor St., SW
Atlanta, GA 30303
Attn. Executive Director
Telephone: (404) 656-3926

Guam

Territorial Board for Professional Engineers, Architects and Land Surveyors
Department of Public Works
Government of Guam
P. O. Box 2950
Agana, Guam 96910
Attn. Chairman
Telephone: 646-8643

Hawaii

State Board of Registration for Professional Engineers, Architects, Land Surveyors and Landscape Architects
P. O. Box 3469 (1010 Richards St.)
Honolulu, HI 96801
Attn. Executive Secretary
Telephone: (808) 548-7637

Idaho

Board of Professional Engineers and Land Surveyors
842 La Cassia Drive
Boise, ID 83705
Attn. Executive Secretary
Telephone: (208) 334-3860

Illinois

Department of Registration and Education
Professional Engineers' Examining Committee
320 West Washington, 3rd Floor
Springfield, IL 62786
Attn. Unit Manager
Telephone: (217) 782-0177

Indiana

State Board of Registration for Professional Engineers and Land Surveyors
Regulated Occupations & Professional Service Bureau
1021 State Office Building
100 N. Senate Avenue
Indianapolis, IN 46204
Attn. Executive Director
Telephone: (317) 232-1840

Iowa

State Board of Engineering Examiners
Capital Complex, 1209 East Court Avenue
Des Moines, IA 50319
Attn. Executive Secretary
Telephone: (515) 281-5602

Kansas

State Board of Technical Professions
214 West Sixth Street, Second Floor
Topeka, KS 66603
Attn. Executive Secretary
Telephone: (913) 296-3053

Kentucky

State Board of Registration for Professional Engineers and Land Surveyors
Rt. 3, 96-5 (Kentucky Engineering Center), Milville Rd.
Frankfort, KY 40601
Attn. Executive Director
Telephone: (502) 564-2680 and 564-2681

Louisiana

State Board of Registration for Professional Engineers and Land Surveyors
1055 St. Charles Avenue, Ste. 415
New Orleans, LA 70130
Attn. Executive Secretary
Telephone: (504) 568-8450

Maine

State Board of Registration for Professional Engineers
State House, Station 92
Augusta, ME 04333
Attn. Secretary
Telephone: (207) 289-3236

Maryland

State Board of Registration for Professional Engineers
501 St. Paul Place, Room 902
Baltimore, MD 21202
Attn. Executive Secretary
Telephone: (301) 659-6322

Massachusetts

State Board of Registration of Professional Engineers and Land Surveyors
Room 1512 Leverett Saltonstall Bldg.
100 Cambridge Street
Boston, MA 02202
Attn. Secretary
Telephone: (617) 727-3055

Michigan

Board of Professional Engineers
P. O. Box 30018 (611 W. Ottawa)
Lansing, MI 48908
Attn. Administrative Secretary
Telephone: (517) 373-3880

Minnesota

State Board of Registration for Architects, Engineers, Land Surveyors and Landscape Architects
Room 162, Metro Square Building
St. Paul, MN 55101
Attn. Executive Secretary
Telephone: (612) 296-2388

Mississippi

State Board of Registration for Professional Engineers and Land Surveyors
P. O. Box 3 (200 South President Street, Ste. 516)
Jackson, MI 39205
Attn. Executive Director
Telephone: (601) 354-7241

Missouri

Board of Architects, Professional Engineers and Land Surveyors
P. O. Box 184 (3523 North Ten Mile Drive)
Jefferson City, MO 65102
Attn. Executive Director
Telephone: (314) 751-2334

Montana

State Board of Professional Engineers and Land Surveyors
Dept. of Commerce, 1424 9th Avenue
Helena, MT 59620-0407
Attn. Administrative Secretary
Telephone: (406) 444-4285

Nebraska

State Board of Examiners for Professional Engineers and Architects
P. O. Box 94751 (301 Centennial Mall, South)
Lincoln, NE 68509
Attn. Executive Director
Telephone: (402) 471-2021 or 471-2407

Nevada

State Board of Registered Professional Engineers and Land Surveyors
1755 East Plumb Lane, Ste. 102
Reno, NV 89502
Attn. Acting Executive Secretary
Telephone: (702) 329-1955

New Hampshire

State Board of Professional Engineers
Storrs St.
Concord, NH 03301
Attn. Executive Secretary
Telephone: (603) 271-2219

New Jersey

State Board of Professional Engineers and Land Surveyors
1100 Raymond Blvd., Room 317
Newark, NJ 07201
Attn. Executive Secretary-Director
Telephone: (201) 648-2660

New Mexico

State Board of Registration for Professional Engineers and Land Surveyors
P. O. Box 4847 (Mays Bldg., Ste. A
440 Cerrillos Rd.)
Santa Fe, NM 87502
Attn. Secretary
Telephone: (505) 827-9940

New York

State Board for Engineering and Land Surveying
The State Education Department
Cultural Education Center
Madison Avenue
Albany, NY 12230
Attn. Executive Secretary
Telephone: (518) 474-3846

North Carolina

Board of Registration for Professional Engineers and Land Surveyors
3620 Six Forks Road
Raleigh, NC 27609
Attn. Executive Secretary
Telephone: (919) 781-9499

North Dakota

State Board of Registration for Professional Engineers and Land Surveyors
P. O. Box 1357 (420 Avenue B East)
Bismarck, ND 58502
Attn. Executive Secretary
Telephone: (701) 258-0786

North Mariana Islands

Board of Professional Licensing
P. O. Box 55 CHRB
Saipan, CM 96950
Attn. Chairman

Ohio

State Board of Registration for Professional Engineers and Surveyors
65 South Front Street, Room 302
Columbus, OH 43266-0314
Attn. Executive Secretary
Telephone: (614) 466-8948

Oklahoma

State Board of Registration for Professional Engineers and Land Surveyors
Oklahoma Engineering Center
Room 120, 201 N. E. 27th Street
Oklahoma City, OK 73105
Attn. Executive Secretary
Telephone: (405) 521-2874

Oregon

State Board of Engineering Examiners
Department of Commerce
403 Labor and Industries Building
Salem, OR 97310
Attn. Executive Secretary
Telephone: (503) 378-4180

Pennsylvania

State Registration Board for Professional Engineers
Mail: P. O. Box 2649
(Transportation & Safety Building, 6th Floor
Commonwealth Avenue & Forester Street)
Harrisburg, PA 17105-2649
Attn. Administrative Secretary
Telephone: (717) 783-7049

Puerto Rico

Board of Examiners of Engineers, Architects, and Surveyors
Box 3271 (Tanca Street, 261, Comer Tetuan)
San Juan, PR 00904
Attn. Director, Examining Boards
Telephone: (809) 722-2121, ext. 268

Rhode Island

State Board of Registration for Professional Engineers and Land Surveyors
Department of Business Regulation
100 North Main Street
Providence, RI 02903
Attn. Administrative Assistant
Telephone: (401) 277-2565

South Carolina

State Board of Registration for Professional Engineers and Land Surveyors
Mail: P. O. Drawer 50408
(2221 Devine Street, Suite 404 29205)
Columbia, SC 29250
Attn. Agency Director
Telephone: (803) 758-2855

South Dakota

State Commission of Engineering and Architectural Examiners
2040 West Main Street, Suite 212
Rapid City, SD 57702-2497
Attn. Executive Secretary
Telephone: (605) 394-2510

Tennessee

State Board of Architectural and Engineering Examiners
546 Doctors' Bldg., 706 Church Street,
Nashville, TN 37219-5322
Attn. Administrator
Telephone: (615) 741-3221

Texas

State Board of Registration for Professional Engineers
Mail: P. O. Drawer 18329
(1917 1H 35 South)
Austin, TX 78760
Attn. Executive Director
Telephone: (512) 475-3141

Utah

Representative Committee for Professional Engineers and Land Surveyors
Division of Registration
Mail: P. O. Box 45802
(160 East 300 South)
Salt Lake City, UT 84145
Attn. Director
Telephone: (801) 530-6628

Vermont

State Board of Registration for Professional Engineers
Division of Licensing and Registration
Pavilion Bldg.
Montpelier, VT 05602
Attn. Executive Secretary
Telephone: (802) 828-2363

Virginia

State Board of Architects, Professional Engineers, Land Surveyors and Certified Landscape Architects
3600 W. Broad Street
Seaboard Bldg., 5th Floor
Richmond, VA 23230-4917
Attn. Assistant Director
Telephone: (804) 257-8512

Virgin Islands

Board for Architects, Engineers and Land Surveyors
Submarine Base
P. O. Box 476
St. Thomas, VI 00801
Attn. Secretary
Telephone: (809) 774-1301

Washington

State Board of Registration for Professional Engineers and Land Surveyors
Mail: P. O. Box 9649
1300 Quince St. (3rd Floor)
Olympia, WA 98504
Attn. Executive Secretary
Telephone: (206) 753-6966

West Virginia

State Board of Registration for Professional Engineers
608 Union Building
Charleston, WV 25301
Attn. Executive Director
Telephone: (304) 348-3554

Wisconsin

State Examining Board of Architects, Professional Engineers, Designers and Land Surveyors
Mail: P. O. Box 8936
(1400 East Washington Avenue)
Madison, WI 53708
Attn. Administrator
Telephone: (608) 266-1397

Wyoming

State Board of Examining Engineers
Herschler Building, Room 4135
Cheyenne, WY 82002
Attn. Executive Secretary
Telephone: (307) 777-6156

Addresses of Canadian Associations of Professional Engineers

Canadian Council of Professional Engineers

Canadian Council of Professional Engineers
Suite 401, 116 Albert Street, Ottawa, Ontario K1P 5G3
Executive Director: Claude Lajeunesse, P. Eng.
Telephone: (613) 232-2474

Provincial Associations (1985)

Association of Professional Engineers, Geologists & Geophysicists of Alberta (APPEGGA) 1010,
One Thornton Court, Edmonton, Alberta T5J 2E7
Telephone: (403) 426-3990

Association of Professional Engineers of British Columbia (APEBC)
2210 West, 12th Avenue, Vancouver, British Columbia V6K 2N6
Attn. Managing Director/Registrar
Telephone: (604) 736-9808

Association of Professional Engineers of Manitoba (APEM)
640-175 Hargrave St., Winnipeg, Manitoba R3C 3R8
Attn. General Manager/Registrar
Telephone: (204) 942-6481

Association of Professional Engineers of New Brunswick (APENB)
123 York Street, Fredericton, New Brunswick E3B 3N6
Attn. Executive Director
Telephone: (506) 454-3296

Association of Professional Engineers of Newfoundland (APEN)
P. O. Box 8414, Postal Station 'A', St. John's, Newfoundland A1B 3N7
Attn. General Manager
Telephone: (709) 753-7714

Association of Professional Engineers, Geologists and Geophysicists of the Northwest Territories (APEGGNWT)
Box 1962, Yellowknife, Northwest Territory X1A 2P5
Attn. Executive Director
Telephone: (403) 920-2477

Association of Professional Engineers of Nova Scotia (APENS)
P. O. Box 129, Halifax, Nova Scotia B3J 2M4
Attn. Executive Secretary/Treas./Registrar
Telephone: (902) 429-2250

Association of Professional Engineers of Ontario (APEO)
1027 Yonge Street, Toronto, Ontario M4W 3E5
Attn. Executive Director
Telephone: (416) 961-1100

Association of Professional Engineers of Prince Edward Island (APEPEI)
P. O. Box 278, Charlottetown, Prince Edward Island C1A 4B1
Attn. Secretary-Registrar
Telephone: (902) 892-9174

Order of Engineers of Quebec/Ordre des Ingenieurs du Quebec (OEQ/O1Q)

2075, rue University, Suite 1100,
Montreal, Quebec H3A 1K8
Attn. Executive Director
Telephone: (514) 845-6141

Association of Professional Engineers
of Saskatchewan (APES)
2255 Thirteenth Avenue, Regina,
Saskatchewan S4P OV6
Attn. Registrar
Telephone: (306) 527-8266

Association of Professional Engineers
of Yukon Territory (APEYT)
P. O. Box 4125, Whitehorse, Yukon
Y1A 3S9
Attn. Registrar
Telephone: (403) 667-6727

Listings courtesy NCEE.

ACCREDITATION BOARD FOR ENGINEERING AND TECHNOLOGY (ABET)

A sound engineering education is a basic ingredient in the development of engineering professionals. The key role of the Accreditation Board for Engineering and Technology (ABET) in the establishment and the maintenance of high standards for engineering programs cannot be overemphasized.

Practically all of the accreditation activities are handled by volunteer engineers from the various technical and professional societies. These engineers give freely of their time and talents. They feel strongly because of their recognition that theirs is a substantial contribution to the future health of the engineering profession . . . our profession.

ABET was established in 1932 under the name Engineers' Council for Professional Development (ECPD) and is primarily responsible for monitoring, evaluating, and certifying the quality of engineering and engineering-related education in colleges and universities in the United States. ABET develops accreditation policies and criteria and conducts a comprehensive program of evaluation of engineering and engineering technology degree programs. Programs that meet the prescribed criteria are granted accreditation status. Note, in particular, that only engineering programs, *not* engineering schools, are subject to the accreditation action.

ABET's legal recognition comes from two sources. First, the U.S. Department of Education periodically reviews ABET's operations and formally recognizes ABET's exclusive jurisdiction for accreditation of engineering, engineering technology, and engineering-related education. Second, state licensing authorities, either by specific statute or by long-standing practice, generally recognize ABET-accredited engineering programs for full educational credit toward satisfaction of state professional engineer licensure requirements.

Evaluation Procedure

Through a process of on-site visitations, ABET provides engineering schools with a means to have their programs formally evaluated against criteria which have been set by the engineering profession. Programs that meet the criteria are awarded a term of accreditation lasting up to six years.

As a result, students in an ABET-accredited engineering or engineering technology program have a definite assurance that their investment in a collegiate engineering education is protected.

The steps in the accreditation process are as follows:

First, a school requests accreditation for a particular *program* in engineering.

Second, a team of volunteer experienced engineering practitioners and/or educators is assembled to visit the school. Equipped with various documents, including a detailed self-evaluation questionnaire completed by the school beforehand, the team conducts an on-site campus visit to consult with administrators, faculty, students, and departmental personnel. The accreditation visit normally encompasses three days, usually starting on a Sunday and finishing on a Tuesday.

The ABET team examines the academic and professional qualifications of the faculty, adequacy of laboratories, equipment, computer facilities, library facilities, and more. The team also looks at the quality of the students' work as evaluated on the basis of interviews with students and the assessment of current examination papers, laboratory work, reports and theses, records, models or equipment constructed by students, and other evidence of the scope of their education.

In addition, the team performs a qualitative and quantitative analysis of program content to ensure that it meets the *minimum* criteria for mathematical foundations, basic sciences, engineering sciences, engineering design and synthesis, and humanities and social sciences which ABET requires to complement the technical education of the student.

Third, the team chairperson prepares a preliminary report with input from the program evaluator. Officers of the Engineering Accreditation Commission (EAC) of ABET review and edit the report and send it to the institution for its "due process" review and comment. This procedure allows the institution to correct any errors of fact or observation. The institution's response is reviewed by the originator of the preliminary report to determine whether changes are warranted.

Fourth, the preliminary report, an analysis of the institution's response, and any other related materials are presented to the full EAC for its review and action. The EAC may grant (or extend) accreditation of a program for a period of up to six years or it may deny accreditation altogether.

ABET Representation

ABET is governed by nineteen "Participating Bodies" that are the major technical and professional engineering societies in the United States. They are represented on the Board of Directors and the Accreditation Commission by individuals who encompass the private, public, and academic sectors of a wide range of engineering disciplines. Along with two Member Bodies, these societies represent over 700,000 individual engineers.

All members of the ABET Board of Directors, of the Engineering Accreditation Commission, and of the visiting evaluation team are volunteers who contribute their time and effort to the accreditation process.

A listing of accredited engineering and technology programs may be obtained by writing to Accreditation Board for Engineering and Technology (ABET), 345 East 47th Street, New York, NY 10017.

RECOMMENDED REFERENCE TEXTS AND STUDY AIDS

Candidates should take into the examination room only those sources of information with which they are most familiar. Texts used in college should furnish excellent review material. Bound notes taken during refresher courses and study aids especially prepared for licensing examinations may be permitted into the examination room. Candidates should check with their board of registration as to conditions required to be met; however, it frequently occurs that a candidate looks for help and advice regarding tests and other aids. The following lists are offered in this light. After looking them over, you may find, in your own estimation, that there are other books which are equal or superior to those listed. Whatever they may be, become familiar with their contents.

In addition to the listings, much help may be obtained from the various technical publications: *Chemical Engineering*; *Mechanical Engineering*; *Heating, Piping and Air Conditioning*; *Machine Design*; *Nucleonics*; *Plant Engineering*; *Power*; *Power Engineering*; and *Research and Development*. It should be noted that some of the data appearing in various sources are updated periodically by their publishers, so that it behooves interested engineers to contact the publishers and obtain new data for evaluation and comparison.

Publication dates are not given in the following list since it has been found from personal experience that any edition is helpful. Exams are developed by experienced professional engineers and textbook solutions are avoided to prevent mere copying. The best way to prepare is to work problems, using the subject listings in Tables 10.1 and 10.2 as guides. Also contact NCEE for a listing of their study guides.

Fundamentals of Engineering

Mathematics for Technician Engineers, Dyball (McGraw-Hill)

Engineering Fundamentals and Problem Solving, Eide (McGraw-Hill)

Introduction to Engineering Calculations, Gottfried (McGraw-Hill)

Engineering Mathematics Handbook, Tuma (McGraw-Hill)

College Chemistry, Frey (Prentice-Hall)

Elementary Fluid Mechanics, Vennard (Wiley)

Heat and Thermodynamics, Zemansky (McGraw-Hill)

Fundamentals of Electrical Engineering, Pumphrey (Prentice-Hall)

Applied Thermodynamics, Faires (Macmillan)

A Programmed Review of Engineering Fundamentals, Baldwin (Reinhold)

Preparing for the Engineer-in-Training Examination, Levinson (Engineering Press, Inc., P. O. Box 1, San Jose CA 95103)

Fundamentals of Engineering (FE) Sample Examination—1983, NCEE

Engineer-in-Training License Review, Newnan (Engineering Press)

Engineering Fundamentals for Professional Engineers' Examinations, Polentz (McGraw-Hill)

Chemical Engineering

Chemical Engineering for Professional Engineers' Examinations, Prabhudesai and Das (McGraw-Hill)

Chemical Engineering—A Review for the PE Exam, Crockett (Wiley)

Unit Operations of Chemical Engineering, McCabe and Smith (McGraw-Hill)

Chemical Engineers' Handbook, Perry et al. (McGraw-Hill)

Introduction to Chemical Engineering Thermodynamics, Smith and Van Ness (McGraw-Hill)

Principles of Heat Transfer, Kreith (Harper & Row)

Chemical Reaction Engineering, Levenspeil (Wiley)

Unit Operations, Brown et al. (Wiley)

Thermodynamic Properties of Steam, Keenan and Keys (Wiley)

Mass Transfer Operations, Treybal (McGraw-Hill)

Manual of Process Calculations, Clarke and Davidson (McGraw-Hill)

Elements of Fractional Distillation, Robinson and Gilliland (McGraw-Hill)

Standard Handbook of Engineering Calculations, Hicks (McGraw-Hill)

Absorption Towers, Morris and Jackson (Butterworth, London)

Chemical Process Principles, Hougen and Watson and Ragatz (Wiley)

Reaction Kinetics for Chemical Engineers, Wales (McGraw-Hill)

Chemical Engineering Kinetics, Smith (McGraw-Hill)

Process Systems Analysis and Control, Coughanower and Koppel (McGraw-Hill)

Process Control, Harriott (McGraw-Hill)

Advanced Engineering Mathematics, Wylie (McGraw-Hill)

Standard Mathematical Tables (The Chemical Rubber Co.)

The Recovery of Vapors, Robinson (Reinhold)

Process Equipment Design, Hesse and Rushton (Wiley)

Civil/Sanitary/Structural Engineering

Timber Construction Manual, American Institute of Timber Construction (Wiley)

Wastewater Engineering—Treatment, Disposal, Reuse, Metcalf & Eddy, Inc. (McGraw-Hill)

Introduction to Highway Engineering, Bateman (Wiley)

Foundation Analysis and Design, Bowles (McGraw-Hill)

Design of Steel Structures, Bresler and Lin (Wiley)

Seepage, Drainage, and Flow Nets, Cedergren (Wiley)

Design of Prestressed Concrete Beams, Connolly (McGraw-Hill)

Handbook of Applied Hydraulics, Davis and Sorenson (McGraw-Hill)

Traffic Flow Theory and Control, Drew (McGraw-Hill)

Standard Handbook of Engineering Calculations, Hicks (McGraw-Hill)

Industrial Water Pollution, Eckenfelder (McGraw-Hill)

Introduction to Transportation Engineering, Hay (Wiley)

Fundamentals of Transportation Engineering, Hennes and Ekse (McGraw-Hill)

Modern Timber Design, Hansen (Wiley)

Structural Concrete, Johnson (McGraw-Hill)

Comprehensive Structural Design Guide, Kurtz (McGraw-Hill)

Concrete Engineering Handbook, LaLonde and Janes (McGraw-Hill)

Handbook of Hydraulics, King and Brater (McGraw-Hill)

Mechanisms of Engineering Structures, Rogers and Causey (Wiley)

Formwork for Concrete Structures, Peurifoy (McGraw-Hill)

Principles of Pavement Design, Yoder (Wiley)

Advanced Reinforced Concrete, Dunham (McGraw-Hill)

Structural Engineering Handbook, Gaylord and Gaylord (McGraw-Hill)

Basic Structural Design, Gerstle (McGraw-Hill)

Linear Framed Structures, Gregory (Wiley)

Water Resources Systems Engineering, Hall and Dracup (McGraw-Hill)

Frame Analysis, Hall and Woodhead (Wiley)

Thin Concrete Shells, Haas (Wiley)

Railroad Engineering, Hay (Wiley)

Standard Handbook for Civil Engineers, Merritt (McGraw-Hill)

Handbook of Engineering Materials, Miner and Seastone (Wiley)

Building Construction Handbook, Merritt (McGraw-Hill)

Field Inspection of Building Construction, McKaig (McGraw-Hill)

Building Failures, McKaig (McGraw-Hill)

Applied Structural Design of Buildings, McKaig (McGraw-Hill)

Highway Engineering, Oglesby and Hewes (Wiley)

Simplified Design of Reinforced Concrete, Parker (Wiley)

Simplified Design of Structural Steel, Parker (Wiley)

Unit Processes of Sanitary Engineering, Rich (Wiley)

Unit Operations of Sanitary Engineering, Rich (Wiley)
Environmental Sanitation, Salvato (Wiley)
Geology in Engineering, Schultz and Cleaves (Wiley)
Soil Mechanics and Engineering, Scott and Schoustra (McGraw-Hill)
Foundation Design and Practice, Seeley (Wiley)
Soil Engineering, Spangler (International Textbook Company)
Soil Mechanics in Engineering Practice, Terzaghi and Peck (Wiley)
Ground Water Hydrology, Todd (Wiley)
Civil Engineering Handbook, Urquhart (McGraw-Hill)
Highway Engineering Handbook, Woods (McGraw-Hill)
Principles of Pavement Design, Yoder (Wiley)

Electrical Engineering

Standard Handbook for Electrical Engineers, Fink and Beatty (McGraw-Hill)

Electrical Engineer's Handbook, Pender and Del Mar (Wiley)

Basic Electrical Engineering, Nau (Wiley)

Circuits, Devices, and Systems, Smith (Wiley)

Electric Machinery, Fitzgerald, Kingsley, and Kusko (McGraw-Hill)

Electric Motors and Their Applications, Lloyd (Wiley)

Direct and Alternating Current Machinery, Rosenblatt and Freidman (McGraw-Hill)

Electric Motor Handbook, Werninck (McGraw-Hill)

An Introduction to Electrical Machines and Transformers, McPherson (Wiley)

Design of Systems and Circuits, Becker and Jensen (McGraw-Hill)

Basic Electric Circuits, Leach (Wiley)

Switchgear and Control Handbook, Smeaton (McGraw-Hill)

Computer Methods in Power System Analysis, Stagg and El-Abiad (McGraw-Hill)

Engineering Circuit Analysis, Hayt (McGraw-Hill)

Mathematical Handbook for Scientists and Engineers, Korn and Korn (McGraw-Hill)

Direct and Alternating Circuits, Oppenheimer, Hess, and Borchers (McGraw-Hill)

Industrial Power Systems Handbook, Beeman (McGraw-Hill)

American Electrician's Handbook, Croft, et al. (McGraw-Hill)

Electrical System Analysis and Design for Industrial Plants, Lazar (McGraw-Hill)

National Electrical Code Handbook, McPartland (McGraw-Hill)

Practical Electricity, McPartland and Novak (McGraw-Hill)

Power Systems Protection: Static Relays, Rao (McGraw-Hill)

Practices and Procedures of Industrial Electrical Design, Roe (McGraw-Hill)

Power System Planning, Sullivan (McGraw-Hill)

Total Energy Management, 2d ed., NEMA

Computer Hardware and Software, Abrams (Addison-Wesley)

A Practical Guide to Microcomputer Applications, Conroy (Wiley)

Practical Applications of Data Communications, Karp (McGraw-Hill)

Application of Computers in Engineering Analysis, Wolberg (McGraw-Hill)

Techniques in Computer Programming, Sherman (Prentice-Hall)

FORTRAN with Engineering Applications, McCracken (Wiley)

Theory and Design of Digital Computer Systems, Levin (Wiley)

Energy Conservation Guide for Industrial Processes. Send inquiries to Northern Division, Naval Facilities Engineering Command, Philadelphia Naval Shipyard, Philadelphia, PA 19112

Mechanical Engineering

Mechanical Vibrations, Church (Wiley)

Machine Design, Creamer (Addison-Wesley)

Design of Machine Elements, Faires (Macmillan)

Mechanics of Materials, Higdon (Wiley)

Kinematics of Machines, Hinkle (Prentice-Hall)

Machinery Handbook, Ohers (Industrial Press)

Design of Machine Elements, Spotts (Prentice-Hall)

Strength of Materials, Willems (McGraw-Hill)

Fluid Mechanics, Binder (Prentice-Hall)

Handbook of Hydraulics, Brater and King (McGraw-Hill)

Fluid Mechanics with Engineering Applications, Daugherty and Franzinia (McGraw-Hill)

Pump Application Engineering, Hicks and Edwards (McGraw-Hill)

Compressible-Fluid Dynamics, Thompson (McGraw-Hill)

Heat and Thermodynamics, Zemansky (McGraw-Hill)

Power Plant Theory and Design, Potter (Wiley)

Internal Combustion Engines, Lichty (McGraw-Hill)

Principles of Turbomachinery, Shepherd (Macmillan)

Engineering Heat Transfer, Welty (Wiley)

Process Heat Transfer, Kern (McGraw-Hill)

Heat Transfer, Holman (McGraw-Hill)

Handbook of Air Conditioning System Design, Carrier Air Conditioning Company (McGraw-Hill)

Plant and Process Ventilation, Hemeon (Industrial Press)

Controlling In-Plant Airborne Contaminants, Constance (Marcel Dekker)

Design of Thermal Systems, Stoecker (McGraw-Hill)

Fan Engineering (Buffalo Forge)

Principles of Refrigeration, Dossat (Wiley)

Heating, Ventilating and Air Conditioning, McQuiston (Wiley)

ASHRAE Guide and Data Book Applications

Nuclear Engineering Handbook, Etherington (McGraw-Hill)

Standard Handbook of Engineering Calculations, Hicks (McGraw-Hill)

Marks Standard Handbook for Mechanical Engineers, Baumeister et al. (McGraw-Hill)

Mechanical Engineer's Handbook, Kent (Wiley)

Cameron Hydraulic Data, Ingersoll-Rand

Cameron Pump Operator's Data, Ingersoll-Rand

Compressed Air Data, Ingersoll-Rand

Economics and Ethics

Engineering Contracts and Specifications, Abbett (Wiley)

Economics, Bach (Prentice-Hall)

Principles of Engineering Economy, Grant and Ireson (Ronald Press)

Legal and Ethical Phases of Engineering, Harding and Canfield (McGraw-Hill)

Other Disciplines

NCEE conducts an ongoing program to solicit examination items (questions) from the memberships of the various national engineering societies as well as from individuals. For a listing of useful specific examination study aids it is suggested you write to the following societies. In doing so, refer to the various subjects covered in each of the Group 2 disciplines in Table 10.2. Also write to your state board to determine if it conducts examinations in your area of interest. Group 2 disciplines examinations are not conducted as often as the Group 1 disciplines.

Aeronautical/aerospace

American Institute of Aeronautics and Astronautics, 1633 Broadway, New York, NY 10019. Telephone: (212) 581-4300.

Agricultural

American Society of Agricultural Engineers, 2950 Niles Road, St. Joseph, MI 49085. Telephone: (616) 429-0300.

Fire Protection

Society of Fire Protection Engineers, 60 Batterymarch Street, Boston, MA 02110.

Industrial

Institute of Industrial Engineers, 25 Technology Park/Atlanta, Norcross, GA 30092. Telephone: (404) 449-0460.

Manufacturing

Society of Manufacturing Engineers, P. O. Box 930, Dearborn, MI 48121. Telephone: (313) 271-1500.

Metallurgical, mining/mineral, petroleum

American Institute of Mining, Metallurgical, and Petroleum Engineers, 345 East 47th Street, New York, NY 10017. Telephone: (212) 705-7695.

Nuclear

American Nuclear Society, 555 N. Kensington, LaGrange Park, IL 60525. Telephone: (312) 352-6611.

Ceramic

National Institute of Ceramic Engineers, 65 Ceramic Drive, Columbus, OH 43214.

General information regarding national engineering societies

American Association of Engineering Societies. A private, nonprofit organization to which all of the major engineering societies belong. AAES advises, represents, and fosters communication among the societies. 345 East 47th Street, New York, NY 10017. Telephone: (212) 705-7840.

SPECIAL STUDY HELPS FOR LICENSING EXAMINATIONS

In preparing for the written examinations, candidates must practice a certain technique so that they will feel "at home" in the examination room. To achieve this end, a number of study helps have been published, written by individuals with years of experience working with license-minded engineers. These publications help focus attention on relevant material and expose the student to a comprehensive selection of actual tested examination problems, solutions, and quick reviews of theory. Actual performance in the examination room has proved the material to be useful, and the approach has been found to provide the examinee with a sense of quick mastery of the subject matter usually found in the exams.

It must be emphasized that the mere knowledge of solutions to selected problems does not constitute knowledge of the entire subject. A previous given problem may be rewritten so that in the examination it is not quickly recognized as one in a question-and-answer book. Basic theory must be thoroughly understood.

In a number of cases the authors have presented more than a mere compilation of problems with solutions, and in so doing, have taken their works out of the class of ordinary question-and-answer books. These authors have presented detailed basic theory, which, if properly digested and augmented by reference to standard texts, should provide the license candidate with the proper background and preparation.

Sample examinations are always an excellent source of help and guidance. Write NCEE for a listing of their study helps. Also, write NSPE for their aforementioned Publication 2201.

NATIONAL COUNCIL OF ENGINEERING EXAMINERS

Purposes and Activities

The National Council of Engineering Examiners (NCEE) is an advisory and coordinating agency established primarily to assist state boards of registration for professional engineers and land surveyors in a cooperative effort for more efficient and uniform administration of state registration laws. It is composed of fifty-five legally constituted state and territorial registration boards divided into four geographical zones, and operates under a president, vice president, and four directors who, with the immediate past president, comprise the board of directors. It has an executive director with offices in Clemson, South Carolina.

Objectives

In the public interest, NCEE provides to licensing boards services which assist in the development and administration of the registration process by promoting:

1. The improvement of registration laws for engineering and land surveying, including the laws, administration, and effectiveness of these laws
2. The uniformity of standards and practices used in engineering and land surveying registration
3. The general acceptance and recognition of comity and reciprocity for engineering and land surveying registration among boards
4. The definition and maintenance of nationally recognized registration qualifications for professional engineers and land surveyors
5. The identification and observation of international professional engineering registration procedures and the maintenance of a liaison with international registration agencies
6. The improvement and uniformity of standards for law enforcement and disciplinary action in engineering and land surveying registration laws and their administration

Aims and Accomplishments

NCEE acts as a clearinghouse for all of its member boards, and is a repository of information on the subject of registration. From its studies of state laws and its comparison of procedures and of requirements for registration, NCEE has compiled valuable data and has aided in bringing about greater uniformity in registration and in gradually raising standards, which it seeks to improve still further. Its contacts provide a medium for the exchange of ideas, enabling individual members to confer, to work, and to cooperate with other member boards in serving the public and the engineering profession. Member boards making full use of NCEE information and facilities have found resultant benefits, often including a saving in time and expense, from use of this central agency. Since national engineering registration is not possible, the only way the states can get any uniformity in action is through such a national organization. NCEE cooperates with all professional engineering societies, serving as a bureau of information for their members.

Through its annual meeting, NCEE provides an opportunity for its members to make personal contacts and learn about others' experiences. Members may take part in the discussion of various subjects (and this they do with much enthusiasm) and secure advice and help from others, if desired. The committee reports and annual-meeting program are carefully prepared, and the proceedings of the meetings are published for the benefit of the members who do not attend, as well as for those who do. Zone meetings are held in connection with the annual meetings and provide a means for members from the four principal sections of the country to compare and discuss with others from their section their particular problems.

General Program

NCEE provides for the continuity of study and work on subjects and procedures of vital and increasing importance to all member boards. The general program of the NCEE consists largely of committee activities, the subjects of which are outlined in the bylaws.

Publications

NCEE publishes *The Registration Bulletin* every three months, issues the proceedings and yearbook shortly after each annual meeting, and compiles a complete digest of state laws governing the practice of engineering and land surveying, and a complete digest of state board practices. In addition, it prepares and distributes both a pamphlet giving general information on legal registration or licensing of professional engineers and containing addresses of registration boards, and the information pamphlet of National Engineering Verification. *The Registration Bulletin* contains concise articles of interest and provides a means of presenting registration information, problems of procedure and administration, facts and figures, preliminary reports, and other matters.

Interstate Registration

Through its committees and officers, NCEE promotes a satisfactory procedure for interstate registration via the Certificate of Verification and provides an effective medium for discussing and developing the solution to this difficult problem of registration.

EXPERIENCE CHECKLIST

The degree of completeness with which the applicant's record is presented can have a telling effect on how much credit the board of examiners will allow. Use this checklist to help write this all-important detail for full value.

1. Has all employment time been accounted for, with no gaps? Account for all time.
2. Was all listed experience gained after age eighteen? Earlier experience is not qualifying.
3. Is all listed experience of an engineering nature? It should be. Diverse experience is given more weight than specialization. All experiences must be attested to.
4. Is all design experience included? Emphasized?
5. Have all areas of the application form been completed? Accounted for? Leave no blanks; nothing should be left to the imagination. Actual nature of duties must be stated so that the board has something concrete to evaluate.
6. Has familiarity with, or responsibility for, at least one project for each employment been indicated? If possible, show magnitude and complexity of the largest project worked on. Describe in detail work done and important decisions made. Description should include name and location of project, capital dollar value if possible, or some other indication of magnitude: tons of steel, cubic yards of concrete, million gallons per day of water pumped, evaporation rate if a steam generator, barrels per day if an oil refinery, etc.
7. Has evidence been shown of experience in its broadest sense, not confined to making drawings or computations?
8. Has progression to more important work and greater responsibility on the job been documented?
9. Has membership in a national engineering society been listed? It will be considered in evaluating eminent experience.
10. Has evidence been indicated of more than a passing interest and initiative during work career? Patents, patents pending, copyrights, publication of technical papers and articles should be listed.
11. Has all responsible charge of engineering work and supervision of employees been shown? Directing the work of others engaged in engineering work or rendering engineering decisions on materials, design, methods of production, operation, or construction programs reflect professional responsibility.
12. Have all college (or other) courses taken for self-improvement and not associated with a degree been listed? Have certificates attesting to the completion of these courses been readied for submittal?
13. Have technical society activities, code committee membership, and the like been listed?

14. Have civic activities and offices (board of education, city council, mayor, etc.) been listed?
15. Has the entire experience and qualification description been presented in an orderly manner for easy evaluation?
16. Have signatures attesting to the character of work of each engagement been obtained?

APPLICATION CHECKLIST

Use this checklist to help your board speed up the processing of your application.

1. Remember that your experience record terminates as of application date, not examination date.
2. Be sure to show place and date of birth.
3. Indicate United States citizenship, or citizenship in foreign country.
4. Indicate name and location of all schools attended.
5. Have two photographs of yourself made: one for application and one to paste on admission card for day of examination. Follow board instructions as to size requirements.
6. Include the required fee with the application.
7. Show courtesy to your references and assure speed in their handling of the reference forms by providing them with addressed and stamped envelopes for mailing directly to the board of examiners.
8. Keep close check on your references. Follow up with telephone call or letter.
9. Write letter of transmittal and include it with your application. Keep carbon copy of all correspondence. Make photocopy of application.
10. Filing application may earmark you for the next examination. If plans change, advise board at time of filing.

APPLICANTS WITH FOREIGN DEGREES

When dealing with applicants with foreign degrees, the primary objectives of boards of registration are:

1. To treat these applicants fairly, using American standards and procedures, with the benefit of doubt being given to assure the public's safety, health, and welfare

2. To make available to the public the knowledge and skills of all professionals who have proven their competency before a board of their peers
3. To require the same degree of excellence and understanding in performance as that required of all practitioners

A keen sense of fairness to fellow professionals prevails in the decisions of all boards of registration. In the final analysis, each board has a basic obligation, duty, and desire to compare each applicant's qualifications with the qualifications required in the provisions of the laws of its own state. Any findings which do not conform with the provisions of the laws or duly adopted rules of its own state may be recognized as not being applicable.

Lack of English Language Proficiency

We must recognize that the lack of English language proficiency, including a limited scientific vocabulary with the accompanying limited ability to understand specifications, codes, or instructions, plus the limited ability to write precise specifications, is serious and unquestionably poses an unacceptable risk to the public health, safety, and welfare. Insisting that a foreign applicant show proof of these talents over and above his or her professional qualifications is a reasonable and essential requirement.

Such an applicant, especially one who has accrued academic qualifications and a majority of practice of engineering in a foreign country, may well be qualified to perform complex assignments at a professional level. An applicant whose academic and practicing qualifications were accrued in a nation in which the metric system is used will require a period of time in working with the English system, where all handbooks, codes, tables, charts, special provisions, and plans use the English system.

Registration grants the registrant the right to practice as an individual with wide latitude for independent judgment. A foreign applicant who possesses these deficiencies is not a safe risk to "turn loose" on the public, notwithstanding acceptable academic qualifications.

Each board of registration establishes its own requirements to measure the proficiency in the use of the English language.

Definition of Terms

Acceptable academic qualifications requires completion of an engineering program adjudged to satisfy the requirements of "graduation from an engineering program approved by the board of record."

Experience credit is credit allowable in many state statutes as part of the total experience requirement for admission to the written examinations. Credit is allowable, in appropriate amounts, for 1) completion of an engineering program adjudged to *not* satisfy the requirements of "graduation from an engineering program approved by the board of record," or 2) graduation from an engineering-related program, or 3) partial completion of an engineering program.

An Engineer-in-Training is a person who holds an EIT certificate issued by a board of registration, having met all statutory requirements and after having passed the 8-hour written examination in the fundamentals of engineering. Enrollment should be regarded as a departure point, not a terminal point in preparing to enter a profession in a foreign country. A foreign applicant may produce credentials that attest to the fact that she or he has gained experience and know-how which clearly demonstrates that that individual has progressed beyond the stage recorded on a diploma. Such a person may be capable of and assigned to perform professional-level assignments as a member of a team or under the supervision of a mature engineer, preferrably licensed as a professional engineer, prior to the time that she or he has developed fluent use of the English language. These facts create a special significance of the 4-year "in training" period of a foreign engineer.

Basic Requirements for Submission of Credentials

A foreign applicant is responsible for having a full transcript of post-secondary academic qualifications submitted by the educational establishment directly to the board of record. This submission must include courses taken, grades received, and the type of entrance requirements achieved.

If the transcript does not suffice, the board will require supplementary application forms that will help the applicant understand that an extra period of time is required before the board can make its initial decision. If translation of the applicant's record is required or if an evaluation of the educational record is required, the foreign applicant should contact the board for guidance and direction in this respect. There are a number of commercial organizations which boards know of who provide a service for this purpose. Addresses of state boards are listed at the beginning of the Appendix.

Credentials from Foreign Schools

1. Certificate of graduation from secondary school (sometimes known as maturity certificate).
2. Student books, official transcripts of subjects or courses with attendance and grades obtained, leaving certificates, *obsolutorium* or other official documents showing exact dates of attendance and subjects pursued year by year in higher and professional study.
3. Diploma and degree; if degree is separate from diploma, submit both.
4. License, if separate license was obtained.

Translation of Credentials

Documents submitted in a language other than English must be accompanied by full, complete, and literal translations of the documents, including both printed and written matter without embellishments or elaborations by the translators. To be acceptable, a translation must be made by a person or establishment properly qualified in the language to be translated.

All translations must contain an affidavit of verification at the end, sworn to by the person making the translation, that the translator has read same after it has been completed; that it is a true and correct translation of the original; that the entire document has been translated and nothing omitted. Failure to comply with this requirement can result in all translations submitted being returned to the applicant for compliance with the requirement. Under no circumstances may the board accept translations by the applicant although in some cases, with board approval, it will be acceptable for the applicant to translate if verified and attested to by a consul general or diplomatic representative duly accredited to the United States and not representing a country behind the "Iron Curtain."

General

The applicant should solicit the names of at least three persons who have personal knowledge of the applicant's schooling, preferably schoolmates now residing in the United States who would be willing to sign affidavits in the applicant's behalf.

Making clear the extent of actual engineering experience within the United States is to the benefit of the applicant. As many samples of work as possible should be presented to the board. Translations must be attested to.

Careful observance of these instructions will greatly facilitate the processing of the application.

SAMPLE EXPERIENCE RECORDS CREDITED BY BOARDS OF EXAMINERS

In evaluating a candidate's experience, his board of examiners must reach its unanimous decision by what is written in his application; it is not enough to list job titles. The application record must give the complete story and must describe the applicant's duties in detail, leaving no doubt on the part of each board member as to the exact nature, character, and extent of the applicant's experience. The number of calendar years of experience does not necessarily equal the number of years of satisfactory "approved" engineering experience.

In general, boards look for experience in positions where an engineering education (or equivalent experience) is an employment requisite and where the work expands the educational experience.

Boards feel that the experience required of an engineering graduate is an internship that serves the purpose of developing and maturing engineering knowledge and judgment. Design work is a very good way of achieving this, but there are many other ways and devices by which it may be accomplished. Boards look for breadth as well as depth.

To help the applicant document his or her experience, several sample records are offered as guides. These case histories have been culled from the author's own experience with hundreds of license-minded engineers over more than 30 years. Boards insist that each person write his or her own record, to reflect individual flavor and provide an insight into the applicant's character and personality. The samples which follow include typical classifications. All names and places have been changed for obvious reasons.

In the first sample, that of an electrical engineer, the write-up disqualified him from sitting for the entire examination because of insufficient experience of a qualifying nature. His board turned him down cold. Sample 2 shows his rewrite and, much to his happy surprise, he was accepted to sit for *all* parts of the examination. Sample 9 is a tested sample of an experience record of outstanding quality. Note the detail given by the applicant.

Sample 1

Education: High school graduate. Graduate of Instituto Industrial "L. A. Huerge," Buenos Aires, Argentina. Associate degree in electrical engineering
Age: 33
Citizen: U.S.A.
Branch of engineering: Electrical
Licensed in another state: No
Member: IEEE, IES
Experience:

6/53 to 3/54	Voss and Voss, electrical and mechanical engineers *Title:* Electrical designer *Duties:* Design of industrial and institutional buildings. Checking vendors' drawings. Field inspections.
4/54 to 5/55	U.S. Army
6/55 to 11/55	Voss and Voss *Duties:* Same duties as previously.
12/55 to 2/57	Gill, Inc., electrical and mechanical engineers *Title:* Electrical designer *Duties:* Design of power plants
3/57 to 7/57	Design Services Associates, mechanical and electrical engineers *Title:* Electrical designer and checker *Duties:* Design of steam power plants. Wiring diagrams, conduit layouts and schedules. Bills of material. Coordinate electrical drawings with other trades.
8/57 to 3/58	Cole and Johnson Co., designers *Title:* Electrical designer and checker *Duties:* Same as for Design Services Associates.
4/58 to 6/63	Hart and Jones, mechanical and electrical engineers *Title:* Electrical project engineer *Duties:* Commercial, industrial and institutional buildings. Supervise and check lighting, power, and communication systems. Write specifications. Supervise and inspect construction sites, coordinate with architects and mechanical trades. Attend owners' and architects' meetings. Perform various administrative duties.

Sample 2

Same individual as for Sample 1. All preliminary items same.
Experience:

6/53 to 3/54	*Duties:* Design and layout of electrical systems and services, selection of equipment and components, knowledge of applicable codes, coordination with other trades, working under minimum supervision of project engineer.

	Typical projects: New York Telephone Company and New Jersey Bell Telephone Company buildings.
4/54 to 5/55	U.S. Army, The Armored School, Fort Knox, Instructor Battalion *Duties:* Technical interpreter-translator (French, Spanish, English). Assignment included extensive work on military technical dictionary. Assisted foreign officers with studies and examinations at the School.
6/55 to 11/55	*Duties:* Design and layout of electrical systems and services, selection of equipment and components, knowledge of applicable codes, coordination with all trades, working under the minimum supervision of the project engineer. *Typical projects:* a. Jewish Medical Center, Baltimore, Md., 6,000-kva load, design completed Nov. 1955. Electrical contract $00,000. b. Bakelite Company, N.Y., 1,000-kva load, design completed Oct. 1955. Electrical contract price $00,000.
12/55 to 2/57	*Duties:* Design and layout of systems in steam power plants. Development of single and control wiring diagrams. Selection of high-voltage switchgear. Distribution of auxiliary systems. Coordination with other trades. Conduit layouts. *Typical projects:* a. Indianapolis Power & Light Company, Harding Street generating station, 100,000 kw. b. White River Generating Station, 75,000 kw.
3/57 to 7/57	*Duties:* Power and lighting design of steam power plants. Development of single and control wiring diagrams. Selection of medium- and high-voltage switchgear. Distribution of auxiliary systems. Conduit layouts and schedules, bills of material. Coordination with other trades. *Typical projects:* a. Public Service Company of Colorado. 110,000-kw steam electric station. b. Pacific Power and Light Company, 45,000 kw.
8/57 to 3/58	*Duties:* Power and lighting design of steam power plants. Development of single line and control wiring diagrams. Selection of medium- and high-voltage switchgear. Distribution of auxiliary systems. Conduit layouts and schedules, bills of material. Coordination with all trades. *Typical projects:* a. Dallas Power & Light Company, 175,000-kw installation. b. Mississippi Power & Light Company, 220,000-kw extension unit.
4/58 to present	*Duties:* On own initiative assumed full resonsibility for engineering design of power, lighting, and low-tension systems, coordinating and maintaining outside contacts on any major project with minimum general supervision, organizing the work and that of other staff members to turn out a complete

set of engineering design plans and specifications, all responsible to Senior Partner. Position requires dealing with clients, the preparation of preliminary reports for proposed projects, working closely with the utility company service engineers, contacting local code authorities for interpretations of various code rules. Troubleshooting during construction, supervising the work of several designers and draftsmen, including the work of two graduate engineers.

Note: The following is a list of major projects in which I had full responsibility for the electrical phases of the project. The only exception is the first assignment, the Chase Manhattan Bank building, whose original design was made by other members of the firm, and for which I have been in full responsible charge since early 1958.

Chase Manhattan Bank, $00,000,000 electrical contract, 22,000 kva. Major decisions: Design responsibility for an extremely elaborate emergency system, using a 500-kw generator located on premises. Completed plans and specifications for a fallout shelter for 18,000 persons. This project included an 800-kw emergency diesel generator.

The air conditioning of Convention Hall, Atlantic City; 3,500 kva. Design included selection of high-voltage multiwinding transformer and associated switchgear, working closely with electric utility engineers.

Tandem Accelerator, New City, 3,500 kva. Design included selection of voltage regulators for stabilization of transit voltages on primary network due to use of very sensitive electronic testing equipment in physics laboratory for nuclear research. Selection and design of very flexible distribution system for high-voltage equipment for use during experiments.

Contributions to engineering knowledge and progress:

Participated in company-sponsored education programs.

Instructor of special courses.

Participated in technical society activities (IEEE and IES), including panel discussions, study groups.

Sample 3

Education: B.S., physics major; 48 graduate credits in physics with emphasis on quantum mechanics, electrodynamics, field theory

Age: 39

Citizen: U.S.A.

Licenses: None other

Member: IEEE

Experience:

1/72 to 9/72 Emerson Radio Co.

Title: Physicist

Duties: Development work on high-frequency and pulsed circuits in components laboratory and radar-receiver laboratory.

9/70 to 9/73	Reeves Instrument Co. *Title:* Physicist *Duties:* Operation of REAC analog computers for solution of guided-missile and aerodynamics problems. Design and construction of auxiliary electronic apparatus.
9/73 to 10/74	NYC Engineering Research Co. *Title:* Research assistant *Duties:* R-f measurements. Theoretical and practical study of detection and amplification of r-f power leading to design of broadband high-accuracy instruments.
10/74 to 2/76	Otis Co. *Title:* Physicist *Duties:* Research on radar systems, airborne bombing, and navigation systems of computers. In charge of group of technicians and junior engineers working on testing and modification of existing systems.
3/76 to 9/78	FRD Co. *Title:* Research physicist *Duties:* Work on information theory, communication systems, and the solid state. Principally, application of Schwinger-Marcuvitz S-matrix and operator formalism to solution of quantum-mechanical problems of electronic states in semiconductors.
9/78 to 9/80	CCNY School of Technology *Title:* Lecturer in physics and electrical engineering *Duties:* Courses taught: laboratory and recitation sections in engineering mechanics, electromagnetics, communications, electronics, atomic and nuclear physics with seminars in relativity and field theory.
9/80 to 3/82	XXXX Corp. *Title:* Research physicist *Duties:* Communications-theory research, patent and paper on new coding and modulation methods. Research in frequency measurements, theory of nonlinear and solid-state devices, noise. Vlf radiating systems, propagation of surface and underseas radio waves.
5/82 to 9/82	Industries Inc. *Title:* Consultant for engineering staff in communications electronics *Duties:* Solution of theoretical and mathematical problems. Design of test procedures and equipment for SSB transmitters.
2/83 to present	Aero Space, Inc. *Title:* Senior staff scientist *Duties:* Responsible for creation of electronics-communications capability in support of contract and proposal efforts. Development of bandwidth compression technique for space radio communications.

Sample 4

Education: B.S., Ch.E.
Age: 35
Member: ASME and AIChE
Experience:

6/72 to 3/76 Air Cool Corp.
Title: Application engineer
Duties: Systems design for air conditioning, ventilation, and exhaust systems. Surveys, selection of equipment, contract checking, control systems, structural design for equipment supports, estimating and reports.

3/76 to 3/78 Combustion, Inc.
Title: Rating engineer
Duties: Design of power-plant steam generators. Selection of fans, blowers, economizers, air heaters, pumps. Combustion calculations, problems in heat transfer, heat and material balances, evaluation.

3/78 to 4/80 Badgett Company, chemical plant and oil refinery design and construction, foreign and domestic.
Title: Assistant to job engineer
Duties: Engineering design of oil refineries in cooperation with project engineers and design groups. Process calculations, equipment selection, piping selection and design. Design coordination with power group.

4/80 to 12/85 Air Fractionation Co., manufacturers of industrial gases.
Title: Staff engineer
Duties: Plant design, selection of equipment, process calculations, rendering of service to the operating plants, study of problems in fluid transport, heat transfer, refrigeration, air conditioning, plant and process water treatment, flowmetering, control systems, plant safety. Special field studies of operating equipment, power services, operating instructions. Vendor contact, requisitioning of equipment, liaison between engineering and operating departments. Initial start-up of processing systems and complete plants.

Contributions to engineering progress and knowledge:
Editor: *Air Conditioning and Elements of Refrigeration*, McGraw-Hill
Technical magazine articles: *Power, Power Engineering, Heating and Ventilating, Operating Engineer, Heating, Piping and Air Conditioning, Chemical Engineering.*

Sample 5

Education: B.S.C.E.
Experience:

9/59 to 7/60 Carnegie Institute of Technology, Staff member, Department of Civil Engineering

	Title: Instructor
	Duties: Instructor in mechanics, structures, and materials testing.
7/60 to 1/63	Military Service, U.S. Army
	Private, Infantry, Basic Training, 5 mo.
	ASTP, University of Illinois, 7 mo.
	OCS, Corps of Engineers, 4 mo.
1/63 to 1/64	Chemical Construction Corp.
	Title: Structural Engineer
	Duties: Engaged in design of support structures for refinery process equipment, including cat crackers, atmospheric, vacuum, polymerization, alkylation, and decoking units.
1/64 to 1/69	*Title:* Structural Engineer
	Duties: Engaged in the supervision of a squad of engineers and draftsmen in the preparation of design drawings for the support of refinery units for specific projects.
1/69 to 7/69	*Title:* Assistant to Division Engineer, Structural Division
	Duties: Engaged in cost appraisals, preparation of specifications for structural steel and concrete design and purchasing. Supervised special structural design projects.
7/69 to 6/78	*Title:* Section Engineer, Structural Division
	Duties: Supervised engineering personnel in the preparation of specifications, design, and drawings for structures, handling equipment, cranes, fireproofing, and special related facilities for refinery and petrochemical units. Supervised and approved the preparation of engineering standards, estimates, and project appraisals related to structural steel.
6/78 to 12/85	*Title:* Division Engineer, Civil Engineering Department
	Duties: Supervised engineering personnel in the preparation of specifications, design, and drawings for structures, foundations, fireproofing, special maintenance equipment for refinery and petrochemical units. Supervised and approved the preparation of soils reports engineering standards, estimates and project appraisals related to Civil Engineering.
12/85 to present	*Title:* Assistant Manager of Design Engineering
	Duties: Supervised Plant Engineering personnel in Civil Engineering Division, General Facilities Division, Piping Design Section, Piping Mechanical Section, Plant Layout Section, and Material Control Section.

Sample 6

Education: B.S. Nuclear Engineer
Age: 49
Citizen: U.S.A.
Experience:

6/76 to 5/79	Atomic Energy Commission, Naval Reactors Division
	Nuclear Power School, Bainbridge, Maryland

Duties: Participated in program of advanced study in nuclear engineering. This course of study was modeled after a Master's degree curriculum in nuclear engineering, with additional, heavily stressed concentration in mechanics, techniques, principles, precepts, and theory of design and engineering of nuclear-fired power plants. Course was very accelerated, with classes 6 days a week, 12 hours a day. Mandatory supervised study of an additional 6 hours a day was required. Although the curriculum exceeded the depth and scope of a typical graduate study program, a degree was not awarded, since the course was a mixture of unclassified university material and highly classified defense information.
Typical Courses: Atomic Physics, Thermodynamics of Reactor Cores. Thermodynamic Properties of Power Plant Materials, Radiochemistry, Metallurgy of Nuclear Fuels, Power Plant Operating Characteristics, Aspects of Nuclear Plant Characteristics, Advanced Engineering Mathematics, Core Design and Engineering Characteristics, Electronics, Electrical Engineering as Applied to Nuclear Power Plants, Quality Assurance and Control Practices, Engineering Assurance, Structural Mechanics of Steam Power Plants. Note: enrollees must be college graduates with engineering experience.

6/79 to 3/80 Atomic Energy Commission, Naval Reactors Division, Nuclear Power Training Prototype Plant, Windsor, Conn.
Duties: Engaged in second portion of study begun at AEC Nuclear Power School. At this school, the student applies the design and engineering knowledge gained during the previous year's schooling to testing and operation of a 20 mw pressurized water reactor power plant. During this instruction period, it was planned for the student engineer to gain experience and expertise in the design and engineering aspects of the power plant systems, each of which he may study and analyze in detail, until all the design and engineering aspects of its inception and fabrication are understood and at the student's command. Additionally, the student must understand the operational aspects of each system separately and of the plant as a whole and qualify as a Plants Operations Supervisor. In this position, the engineer must supervise the operation of the power plant and the activities of the plant's twenty-member operational, repair, and testing team. Finally, a detailed knowledge of the design conditions and operating conditions and characteristics, with specific knowledge of the architectural engineering that went into development, must be within each graduate's grasp.
Typical Areas of Study:

A. For each power plant system and reactor containment:
   1. Structural engineering considerations
   2. Stress analysis
   3. Environmental engineering considerations

4. Electrical and control engineering considerations
5. Design characteristics
6. Quality control criteria and techniques
7. Seismic criteria
8. Mechanical engineering considerations

B. Power plant operating characteristics
C. Core physics and operating characteristics
D. Casualty response
E. Design basis accident criteria and defense
F. Radiological controls as applied to:
   1. Health physics
   2. Shielding
   3. Exposure control
   4. Irradiation control and radiation effects on plant materials
G. Standard engineering and design tools

4/80 to 9/82 — U.S. Navy, Engineer

Duties: As engineer, was responsible for the administration of a 40-member engineering department which operated a 100 mw nuclear steam propulsion plant. Second, the department engineered and supervised an overhaul of nuclear submarines. Such overhauls consisted of nuclear refueling, ship modifications utilizing both newly selected equipment and existing modified equipment, and repair. Finally, specialized in precriticality and acceptance testing which was performed before Navy acceptance of the overhauled ship. Specific tasks included design and engineering of systems for nuclear refueling operations and support and modifications of existing equipment and principles. Specific attention was devoted to special resting and experimenting in development of quality control techniques.

Specific Accomplishments: Developed a special preoverhaul testing program, aimed at revealing all areas of weakness in a power plant which should be addressed during subsequent overhaul. Overhaul, refueling, modification, modernization, and repair of USS *H. C. Marshall* SSBN654 in 25 percent of allotted time.

10/82 to 10/84 — General Electric Company, Knolls Atomic Power Laboratory, Schenectady, New York

Duties: Employed jointly as Plant Engineer and Training Engineer. I was engaged in a multifaceted research and development program and power plant operator training program for the AEC, Naval Reactors Branch. Primary responsibility centered on the research and development of advanced, highly enriched, very compact uranium/plutonium cores for U.S. Navy utilization in the late 1980s and early 1990s. Project included 50 engineers, technicians, and designers. Under moderate supervision, I administered 64 test engineers/instructors in a test program and design effort toward reactor core physics and geometry development. In addition, I conducted, using the above engineers, the instruction of student engineers in the Nuclear Power Prototype Training Program which I described in my 6/76 to 3/77 experience. Students under my control numbered about 150 and were both civilian and military.

The subjects which I specifically taught were:

1. Radiological Controls
2. Reactor plant operating characteristics
3. Quality Control Techniques
4. Thermodynamics

Specific Accomplishments: Development of scheduling techniques which minimized interfacing problems of training and testing utilization of 100 mw test reactor at prototype, which also provided means of direct test information retrieval following conduct of tests. Advanced reactor core design, control rod/burnable poison research, natural circulation primary plant research, and Core Safety analysis techniques development.

10/84 to Present — SWES Corporation, New York

Duties: From 11/80 to 5/81 I served on the ABC Nuclear Power Station, nuclear power plant project as Assistant Nuclear Steam System Supplier (NSSS) Coordinator, Reactor Coolant Systems Engineer and Reactor Coolant Drain System Engineer. In addition, I had the responsibility of supervising all engineering work within the reactor containment structure. In these capacities, I was in charge of four subordinate engineers. Under moderate supervision, I performed the following: Design and engineering of mechanical nuclear systems for pressurized water reactor (PWR) power plant; coordination and interfacing of mechanical nuclear, structural, environmental, electrical, and controls engineering within the reactor containment; selection of equipment and components for nuclear systems; coordination of intercompany relations between client, Company, and NSSS in the area of the nuclear systems. During this time, I performed for the company and particularly for the plant, a code and code application study of the ASME code. The major objective of this study was the selection of proper code classes for NSSS systems.

In 5/74, I was promoted to the minimal supervision position of full NSSS Coordinator for the ABC Nuclear Power Station, units 1 and 2 Project. With seven subordinate engineers, I performed the same functions as previously described for the earlier project. Responsibilities included all emergency core cooling systems, in addition to the reactor coolant system, and chemical and volume control systems.

Summary of Job Functions—All Engagements

1. Engineering calculations
2. Design practices
3. Design criteria

4. Site and plant arrangements
5. Feasibility studies
6. Systems analysis and control reports
7. Specifications—nuclear equipment
8. Vendor liaison
9. Evaluations and recommendations
10. Consultations
11. Coordination of design, engineering, and construction
12. Compliance with client requirements
13. Interdisciplinary interfacing
14. Technical supervisor
15. Schedule adherence
16. Machine Design
17. Schedule monitoring
18. Drawing approval
19. Research and development
20. Contract administration
21. Loop cooling systems
22. Reactor containment
23. Control systems design

Sample 7

Education: Engineering degree (B.S. M.E.)
Citizen: U.S.A.
Member: ASME, IEEE
Previous registration: Connecticut, Massachusetts, Rhode Island
Experience:

2/47 to 4/57 — L and L Mfg. Co., textile machinery
Title: Assistant mechanical engineer
Duties: Time studies, reports, investigations, supervision in electrical and mechanical development. Process development.

4/57 to 2/64 — Hathaway Mfg. Co., textiles
Title: Assistant plant engineer
Duties: Complete plant rehabilitation, acquisition of new properties, new buildings, new lighting installations, new steam plant, electrical distribution and air conditioning (720 tons of refrigeration). Total improvement during the period of engagement was $0,000,000. Total staff 119, plus 100 persons at various times working for outside contractors.

2/64 to 6/84 — C & C, Inc., textiles
Title: Plant engineer
Duties: In charge of maintenance, construction, and engineering. Consultant in engineering for other company-owned plants. Some of the projects undertaken: design and installation of evaporative cooling air-conditioning system, modernization of boiler plant, including forced draft fans, feedwater pumping, combustion controls. Design and installation of two

new heating plants and systems (hot water 20 million Btu). New fluorescent-lighting systems, power-factor-correction programs, short-circuit study, electrical-distribution system consisting of 3,500 kva of transformers, bus duct, switchgear, armored cable. Work force included twelve supervisors and four draftspersons under my direction and responsibility.
Private consultant: Church electrical systems, lighting, electrical heating, public-address systems.

Sample 8

Education: B.S. M.E.
Branch of engineering: Mechanical
No other licenses, no membership in technical societies
Experience:

4/60 to 4/65 — Monroe Heating Co., heating, ventilating, and air conditioning
Title: Heating and air-conditioning engineer
Duties: Supervision of heating-systems design, plan, and specification preparation. Selection of equipment, load calculations, air-distribution design, troubleshooting in field. Tons of refrigeration: 5 to 100.

3/63 to 4/65 — Rochester Heating & Plumbing Co.
Title: Heating and air-conditioning engineer
Duties: Estimating, preparation of working drawings and details for plumbing and air conditioning.

4/65 to 8/85 — Roe & Associates
Title: Heating and air-conditioning engineer
Duties: Complete design and supervision of heating systems, air conditioning, refrigeration.

Sample 9

Member: American Society of Civil Engineers—Fellow, 1959; American Management Association—Member, 1967
Experience:

10/53 to 2/54 — University of Florida, Gainesville, Florida
Title: Instructor in hydraulics
Type of Work: Teaching full-time, at college level, in an accredited school of engineering.
Reported to: Dean———, PE
Old Westbury, Long Island, NY 11568
Duties and Degree of responsibility: Instructed at university level, U.S. Army classes. Specialized Training Program for students in hydraulic theory. Instructed undergraduate and graduate laboratory courses in Spanish to Latin-American students. Subjects Included: Hydrostatics, hydrodynamics, flow in conduits, branch piping, pumps and pumping, hydraulic gradient.

2/54 to 7/56 *Title:* Technical observer (radar) and airborne radio and radar maintenance and operations officer.
Position: Instructor: Physics, applied sciences.

7/56 to 4/62 SWECOM, Boston, Mass.
Title: Senior Designer
Reported to: ———, PE
Duties and degree of responsibility: Responsible for the design and checking of reinforced concrete for industrial buildings and power plants, including: spread, continuous and raft-type foundations on soil and on piles; beam, column, and girder type steel construction; floor slabs; turbine supports and their foundations designed to withstand earthquake acceleration and vibrations; heavy vibrating machine foundations; retaining walls, screen well structures; cooling water intake and discharge structures; circular and rectangular coffer dams; steel sheet and timber piling bulkheads, reinforced concrete tanks; pressure tunnels; water pumping stations; pipe anchors for precast concrete pipe lines on soil and on piles; flat slab-on-grade design; electrical high tension transmission tower foundations; concrete chimneys and their foundations on soil and on piles; rigid frame analysis for different type structures and structural steel design. Job functions included: engineering calculations, design criteria, design practices, field surveys, field sketches, field trouble reports.
Major assignments during this engagement were:

- The Hartford Electric Light Company: Alterations to South Meadow Station for installing mercury arc boiler. Steel framework design. Gulf States Utilities Company, Baton Rouge, LA: Two 40,000-kw additions to power station. Turbine support and waterfront work.
- Savannah Electric and Power Company: Extension to Riverside Station. Foundation design for wood pile raft type foundations

4/62 to 1/64 CMK del Peru, Lima, Peru
Title: Assistant to Chief Design Engineer
Reported to: J.E.S., PE, Sacramento, CA
Duties and degree of responsibility: Structural Design and Engineering Administration. I worked in close association with Peruvian government engineers, translating technical documents and reports from English to Spanish. I performed administrative duties with regard to personnel and assisted in the planning and assignment work as well as the supervision of structural design and some hydraulic design.
Some typical major assignments included:

- Rio Quiroz Irrigation Project: Irrigation design for a flow of 2,100 cfs or 180 million cf per day.
- San Lorenzo Storage Dam: Preliminary design for an earth-storage dam 200 ft high designed to store 239,000 acre-ft of water.

Job functions: Field surveys, proposals, reports, possibility studies, specifications, consultations, project management, contract documents, coordination of design engineering and field construction relations, compliance with client requirements.

1/64 to 9/65 MEECO, Boston, Mass.

Title: Structural and hydraulic engineer

Duties and degree of responsibility: Responsible for design of timber, concrete, and steel structures in connection with sanitary projects. Assisted the chief structural engineer in selection of design approaches, and selection of materials of construction. Maintained design requirements and supervised scheduling of typical following projects:

- Allegheny County Sewerage Authority (Pennsylvania): five story incinerator building. Steel and reinforced concrete design.
- Limestone Air Base, Limestone, Maine: Water treatment plant. Designed reinforced concrete tanks and building foundations. Requirements for this position: a strong technical background in the areas covered, and administrative ability to act in the best interests of the client and the company. A major asset is the ability to coordinate and organize the work of others to achieve a common goal within the framework of contractual obligations and company policies and interests.

9/65 to 6/71 SWECOM, Boston, Mass.

Title: Engineer—Structural Division

Duties and responsibilities: Assisted in the preparation of specifications, purchase of structural materials, supervision of structural design and preparation of estimates.

The following is a partial list of major assignments completed:

- Virginia Electric and Power Company, Bremo Power Station. Responsible for the preparation of an engineering study, drawings, estimates of construction costs and recommendations for type of construction of a new ash disposal area to handle approximately 300,000 kw of installed generating capacity together with final design for construction.
- Inland Steel Company, Proposed Station No. 4-C. Project Engineer in charge of the preparation of a general design specification and inquiry documents to be used in the selection of an engineering-construction firm to design and construct a 40,000-kW turbine generator and two 350,000-lb/h boilers.
- Standard Coated Products, Inc., Buchanan, New York. In charge of all work associated with the addition of a new floor system inside existing building, a new exterior shipping platform, and recommended alterations work required for the replacement of existing floor systems.

6/71 to 6/74 Wasco, Boston, Mass.
Title: Structural Engineer
Reported to: B.M.J., PE
Duties and responsibilities: Responsible for all engineering work, including preparation of engineering criteria, supervision and development of structural engineering and design, also assisted in conceptual layout work and the preparation of estimates for major plant work.
Following is a partial list of major assignments completed:

- Virginia Electric and Power Company, Mount Storm Power Station, Units 1 and 2, Mt. Storm, West Virginia.
- Two 565,000-kW turbine generators and two pulverized coal-fired 3,785,000-lb/hr reheat boilers operating at 2620 psig and 1000/1000°F. This station is a new site and includes the design and construction of a 135-ft-high rockfill dam to impound a 47,000-acre-ft reservoir for cooling water purposes. A 16-mile access railroad over the Allegheny Mountains for access from Bayard, W. Virginia, was designed and constructed. Extensive service facilities, structures, and coal-handling system are also included.

Significant engineering decisions made on this assignment:

1. Type of foundation to be used
2. Bearing value of foundation material
3. Implementation of new AISC design standard for steel design
4. Design criteria and type of construction for combined chimney and fly ash storage silo
5. Type of material and protective coating specifications to protect submerged structures from effect of highly acid (pH 4.2) cooling pond
6. Design criteria, type of material, and construction procedures for a pavement capable of withstanding load of 100-ton coal trucks with specifications for 13-in-thick concrete pavements and 27-in-thick asphalt concrete pavements
7. Development of anchorage system into shale material to handle uplift wind loads on building columns
8. Railroad layout and procedures for handling, dumping, and storing coal cars using radio control systems
9. Specifications for the design and construction of 9-ft-diameter special prestressed embedded cylinder concrete pressure pipe to withstand highly acid circulating water along new cold-applied epoxy protective coatings
10. Use of reinforced concrete forced draft air ducts to attenuate serious noise control problem

6/74 to 4/76 SWECOM, Boston, Mass.
Title: Proposal Engineer
Reported to: W.S.R., PE
Duties and responsibilities: In responsible charge of activities

related to the handling of proposal and contract work associated with procurement of new engineering and construction projects. Work included development and assembly of brochures and similar material suitable for submission to prospective clients to inform them of company capabilities, background, and experience. This also included evaluation of services and inquiries, determination of company interest, and assembly of formal proposals. Most of this assignment was in the fossil and hydroelectric power plant fields.

I was also responsible for the following:

1. Preparation of proposals covering scheduling and coordination of the efforts of individuals assigned to the work from different departments, together with the development of contractual phases of the proposal and the final writing, organization, compilation, and editing of all work in the proposal.
2. Participation in the development of proposal prices and fees along with presentation of proposals to clients.
3. Development of formal contracts outlining the contractual basis for performing work and the participation in discussions with the new client leading to the finalization of the contract language.

4/76 to 8/79 SWECOM, Boston, Mass.

Title: Manager—Production Scheduling and Review Department

Reported to: J.R.C., PE

Duties and responsibilities: Set up a new department within the company charged with the authority for developing and implementing company policy within the fields of planning and scheduling, manpower resource allocation, and project reporting systems for major fossil, hydroelectric, and nuclear power plants. Responsible for the implementation, improvement, and refinement of current procedures; initiation of studies and exploration of techniques and practices in the field of network based management systems; investigation, recommendations for solutions, for project planning and scheduling problems.

Significant achievements with which I was credited and which occurred in this engagement included:

1. Introduction of CPM (critical path method of planning networks) as the official company procedure for planning and scheduling projects.
2. Conversion of the company computer-based network program from NASA PERT "C" to the IBM PMS-360.
3. The development of a formal course of instruction for network-based planning and control which is conducted for members of the client organizations as well as for the members of the engineering, design, and construction forces of the company.

4. Development of a network-based planning and control system, including features such as:
   a. Project breakdown structures
   b. Summary, intermediate, and detailed levels of networks, showing integrated engineering, design, and construction activities
   c. Work analysis sheets
5. Decision to use activity-on-arrow networking instead of activity-on-node as company standard
6. Development of a computer-based work force allocation and leveling program (RAMRODS) linked to the PMS-360 time program
7. Hiring, training, assigning, and supervision of Planning Engineers engaged in these activities resulting in increasing the department size 400 percent to 35 employees.
8. Planning manhour estimates for projects and cost control.
9. Development and administration of internal department budgets for improvements in techniques and procedures.

Important areas of activity which require the personal attention of the Design Manager in both technical and administrative areas are:

1. Execution of corrective measures based on investigation and recommendations for solution to technical, operational, and development problems.
2. Institution of uniform approaches to the solution of technical problems and procedures common to the various design offices and the means for implementing their use and measuring their effectiveness.
3. Overall coordination of Design Division activities with other divisions within the engineering department and other departments' work which interfere with Design Operations. Specific attention is given to planning and scheduling operations where the management of the Engineering/Design/Construction division interfaces are involved.
4. Implementation of a long-range work force development program through intensive internal training and feedback program.
5. Establishment of the Conceptual Design Group in March 1971 with a staff of 25. In connection with the Conceptual Engineering Group, the company formally set out to use uniform approaches in engineering, layout, and designing of all projects and operations of the Conceptual Group. The management of all Project Design Engineering now became the implementer of the Design Network-Based Management System, which comprises the following elements:
   a. Project breakdown structures
   b. Level 1 and 2 networks
   c. Level 3 schedules
   d. Work analysis sheets

8/79 to 1/82 SWECOM, Boston, Mass.
Title: Design Manager
Reported to: W.P.A., PE
Duties and responsibilities: In charge of all personnel and physical plant engaged in the preparation of all technical design and construction drawings for power industry work performed in the Boston office of the company, including New York and New Jersey offices of the company. Accountabilities included planning, organizing, staffing, directing, and controlling activities in both technical and administrative areas of the Design Division for over 1300 employees in engineering. Activities included the implementation, improvement, and refinement of the current operating procedures; introduction of advanced technical design procedures; management techniques and practices.
I was placed in charge of the following:

1. Network-based planning and scheduling system for fossil and nuclear power plants.
2. Terminal-oriented on-line interactive and batch-type computer-based information system to handle administrative and technical operations.
   - A. Administrative systems include employee information, production management, and work flow computer program development and implementation.
   - B. Technical systems include the development and implementation of:
     1. Electrical cable schedule and raceway programs
     2. Application of STRESS and STRUDL programs
     3. Application of pipe stress and pipe support design programs
     4. Network-based work force loadings
        The integrated time/work force/cost system uses the terminal-oriented on-line planning and control system to furnish the needed management reports. The query language is RAMIS
        The following reports are being generated by this system:
        - a. Progress or work accomplished on a planned versus actual basis
        - b. Manhour usage on a planned versus actual basis
        - c. Indices: progress accomplished—planned plus actual; workhour usage, planned versus actual; performance progress index plus workhour index; trend curves

1/82 to 9/82 SWECOM, Boston, Mass.
Title: Assistant Engineering Manager
Reported to: R.M.I., PE

Duties and responsibilities: Reported to the Engineering Manager-Vice President and acted for him in the management of activities of the following divisions in the Power Group:

1. Design Division, Boston, New York, and New Jersey Offices
   Position assumed the accountabilities and activities assigned to the Design Manager as described in 8/79 to 1/82 position.
2. Field Engineering and Design Offices
   Management of all activities involved in planning, organizing, staffing, and directing the establishment of field engineering and design offices for headquarters engineering and design personnel assigned to those offices.
3. Geotechnical and Structural Divisions
   Assisted in formalizing and implementing the Engineering Manager's policies affecting these divisions; directed the activities involved in performance appraisal and salary administration; performed final review of prospective employees and review and approval of salary administration; gave advice and assistance in resolving administrative problems confronting these division heads; reviewed and approved work order budgets and monitored performance and expenditures associated with them.

Company Education Programs: Contributed to engineering knowledge and progress by active participation in the sponsorship of the following company programs for the development of human resources:

1. Provided for and encouraged the conduct of the company continuing education program through complete in-house courses to upgrade development of engineering personnel.
2. Provided courses, guidance, study aids, and financial aid for the Professional Engineers Registration Program.
3. Sponsored the Tuition Assistance Program, which covers 100% assistance for job-related courses and academic degrees in engineering and management.
4. Sponsored the associate degree program, whereby graduates of community colleges with associate degrees from local institutions of learning are selected to participate in the company program of continuing their engineering education at night while gainfully employed during the day.
5. Arranged to enable a selected group of engineering graduates each year to participate in the career development program, which includes training assignments within the company's engineering design, construction, and various other segments of the organization. The program is intended to provide a broad background of experience to the

individual selected through the medium of actual productive work in these activities.

Scholastic Honors:

University of Florida—Tau Beta Pi
University of Florida—Sigma Tau
University of Florida—Phi Kappa Phi

RESPONSIBLE CHARGE

"Responsible charge" means a degree of competence and accountability gained by technical education and experience of a grade and character sufficient to qualify an individual to personally and independently engage in and be entrusted with the work involved in the practice of engineering.

"Responsible charge" may mean that you have "directed the work of others engaged in engineering work." Or it may mean that you have "decided on materials, design, methods of production, operation or construction programs, and have familiarity with related functions such as cost accounting, maintenance and repairs, and installation."

"Responsible charge" includes the independent control and direction—by the use of initiative, skill, and independent judgment—of the investigation or design of professional engineering work or the supervision of such projects. It means charge "of work" and/or "of people." Work assigned by a supervisor (preferably a registered professional engineer) but requiring the supervisor's direct and continued supervision may not be classified as in "responsible charge" on the application form. Boards look for "healthy" progress on the job. College graduates who start out on the drafting board must show how they have progressed to the point where they are independently applying basic engineering principles in their everyday work in the engineering design office or on the job site or in the field.

The developing engineering registration candidate who is building up an acceptable experience record for board evaluation should become involved in the span of control and in engineering decisions for which the candidate's licensed supervisor is responsible. Of course the candidate will not have the final say, but he or she can gain the much needed experience that will qualify him or her at some future date.

This practical engineering experience should stretch over a 4- to 8-year period (depending upon the candidate's education) during which the aspirant contributes to the engineering thinking on the job, such as making minor decisions on design features, specifications, methods of procedure, and the like. The candidate does not have to be responsible for the final design, but she or he does have to contribute ideas based on experience and engineering fundamentals.

The Ideal Supervisor

The ideal supervisor should be a licensed professional engineer with experience in his or her chosen field. The supervisor's span of control directly relates to the degree of control a professional engineer is required to maintain while exercising independent direction of engineering work and to the engineering decisions which only the professional engineer can make.

The *span of control* necessary to be in responsible charge should be such that the engineer:

1. Personally makes the decision that could affect the health, safety, and welfare of the public. A later review and approval of the decisions made by subordinates is not sufficient.
2. Judges and qualifications of technical specialists and the validity and applicability of their recommendations before such recommendations are incorporated in the work or design.

Engineering Decisions

The term "responsible charge" relates to engineering decisions within the purview of registration laws; it does not refer to management control in the hierarchy of professional engineers except insofar as each of the individuals in the hierarchy exercises independent engineering judgment and thus responsible charge. It does not refer to such administrative and personnel management functions as accounting, labor relations, performance standards, marketing of services, and goal setting. While an engineer may also have such duties, this fact should not enhance or decrease one's status of being in responsible charge of the work. The phrase does not refer to the account of legal liability.

Engineering decisions which can be made by the engineer in responsible charge are generally those which must be made at the project level and higher. Examples of assignments where such decisions are commonly made are project design engineer and resident engineer. Such decisions, concerning any permanent or temporary work, the failure of which creates a hazard to life, health, property, or public welfare, include:

1. The selection of engineering alternatives to be investigated and the comparison of alternatives for engineering works

2. The selection or development of design standards or methods and materials to be used
3. The selection or development of techniques or methods of testing to be used in evaluating materials or completed works, either new or existing
4. The review and evaluation of contractor's construction methods or controls to be used and the evaluation of test results, materials, and workmanship insofar as they affect the character and integrity of the completed work
5. The development and control of operating and maintenance procedures

Responsible Charge Criteria

As a test to evaluate whether an engineer is in responsible charge, the following must be considered:

1. The professional engineer who signs engineering documents must be capable of answering questions asked by equally qualified engineers.
2. These questions would be relevant to the engineering decisions made during the individual's participation in the project, and in sufficient detail to leave little question as to the engineer's technical knowledge of the work performed.

It is not necessary to defend decisions as in an adversary situation, but only to demonstrate that the individual in responsible charge made them and possessed sufficient knowledge of the project to make them.

Examples of questions to be answered by the engineer are criteria for design, methods of analysis, methods of manufacture and construction, selection of materials and systems, economics of alternate solutions, and environmental consequences. The individual should be able to define clearly the span of control and how it is exercised both within the organization and geographically and to demonstrate that the engineer is answerable within the said span of control.

SUBJECT LISTINGS OF NSPE PUBLICATIONS (ABRIDGED)

For a complete listing write the National Society of Professional Engineers (NSPE), P.O. Box 35023, Washington, DC 20013.

Pub. No.	Title
2201	*Bibliography on Professional Engineers' Examination Questions*
2202	*Model Law*
2213	*State Engineering Registration Boards* (directory)
2214	*Engineering Registration and the Fundamentals Examination (NCEE)* (folder)
2215	*State-by-State Summary of the Requirements for Engineering Registration*
2216	*Engineering Registration Now: A Guide to Engineering Professionalism*
2217	*State-by-State Summary of the Engineering Corporate Practice Laws*
2218	*State-by-State Guide to Fundamentals of Engineering and Principles and Practice of Engineering Review/Refresher Courses*
1001	*Accredited Engineering Curricula*
1107	*NSPE Ethics Reference Guide*
2601	*NICET Certification Programs Booklet*
1809-A	*Engineers in Industry—Take Charge of Your Future*

DEFINITIONS OF ENGINEERING DISCIPLINES

Practicing engineers within the workplace may be engaged in one or more of the following major functions of engineering: teaching, research, development, design, construction, production, operation and maintenance, application and sales, and industrial management or administration. So long as the application of engineering sciences and an engineering degree are required to hold the job, these functions take on real meaning and may be considered as qualifying experience. If the public welfare or the safeguarding of life, health, and property is involved, these functions apply in whole or in part to the approved classifications of engineering, whether or not they are included in the classification descriptions which follow. The techniques included in the description of each classification, on which applicants for licensure in that classification will be examined, are intended here to be mostly descriptive. The classification should not be confined to the techniques listed. These definitions can be helpful tools in writing up an engineer's experience record, since they closely match job descriptions used by industry and government. A person's experience in engineering may cut across the lines of discipline demarcation shown herein, in which case this listing may prove helpful as a checklist.

1. *Aerospace Engineering.* The art of designing aircraft and aircraft components, missiles, space vehicles, guidance systems, and rockets, and guiding the technical phases of their manufacture and operation.

2. *Agricultural Engineering.* Concerns the design, construction, and use of specialized equipment, machines, structures, and materials relating to the agricultural industry and economy. It requires knowledge of the engineering sciences relating to physical properties and biological variables of foods and fibers; atmospheric phenomena as they are related to agricultural operations; soil dynamics as related to traction, tillage, and plant-soil-water relationships; and human factors relative to safe design and use of agricultural machines. The safe and proper application and use of agricultural chemicals and their effect on the environment are also concerns of the agricultural engineers.

3. *Ceramic Engineering.* Includes the preparation of nonmetallic minerals from raw materials; forming by presses, molds, and wheels; firing in kilns, ovens, and furnaces; and applications to industrial and domestic uses. Ceramic engineering includes processing and manufacturing of abrasives, glassware, building materials, cements, refractories, enamels, white wares, and the like.

4. *Chemical Engineering.* The application of the principles of the physical sciences, together with the principles of economics and human relations, to fields that pertain directly to processes and process equipment in which material is treated to effect a change in state, energy content, or composition. The chemical engineer deals with problems arising in manufacturing processes involving both chemical reactions and the structures, equipment, and machinery necessary for the proper control of a chemical reaction. These processes are usually resolved into a coordinated series of unit physical operations and unit chemical processes. The chemical engineer is concerned with the research, design, production, operational, organizational, and economic aspects of the above.

5. *Civil Engineering.* The most diverse branch of engineering, it includes all engineers engaged in the planning, designing, construction, engineering economics and maintenance of bridges, buildings, waterways, dams, railroads, airport terminals, pipelines, highways, sanitary systems, foundations, hydroelectric installations, irrigation systems, and similar systems and structures of modern civilization. Recently civil engineering has been broadened to include community planning and, in addition to traditional surveying and mapping, has encompassed photogrammetric methods.

6. *Control System Engineering.* Concerns the science of instrumentation and automatic control of dynamic processes and requires the

ability to apply this knowledge to the planning, development, operation, and evaluation of systems of control so as to insure the safety and practical operability of such processes.

7. *Corrosion Engineering*. Concerns the environmental corrosion behavior of materials and requires the ability to apply this knowledge by recommending procedures for control, protection, and cost effectiveness resulting from the investigation of corrosion causes or theoretical reactions.

8. *Electrical Engineering*. Involves the planning, design, construction, testing, maintenance, and operation of electrical power machinery including all devices used in the generation, transmission, distribution, measurement, and utilization of this form of energy. The electrical engineer must be competent in the field of electronics to deal with circuits, vacuum tubes, transistors, antennae, wave propagation, and communications including radio, television, radio-telephone, and the like, or qualified in the area of illumination to deal with light sources measurement, interior and exterior applications, and photoelectric systems. The electrical engineer may be expert and qualified in sound systems and their applications.

9. *Fire Protection Engineering*. Concerns the safeguarding of life and property from fire and fire-related hazards and requires the ability to apply this knowledge to the identification, evaluation, correction, or prevention of present or potential fire and fire-related panic hazards in buildings, groups of buildings, or communities and to recommend the arrangement and use of fire-resistant building materials and fire detection and extinguishing systems, devices, and apparatus in order to protect life and property.

10. *Highway Engineering*. The highway engineer must possess a knowledge of highway and street engineering including planning, designing, economic studies, and construction of the ways, roads, and streets over which the vehicular traffic of the state travels. The highway engineer must possess knowledge of the methods of determining traffic volumes, vehicle characteristics, driver characteristics, and traffic capacities of highways and streets. In addition, this type of engineer must understand surveying, characteristics of soils, methods for increasing the supporting power of soils, pavement design, selection of materials used for pavements, and drainage of ground and surface water, and have a knowledge of construction practice and constuction machinery used in building roads and streets. By using all factors included above, the highway engineer must be able to develop an economical plan for ways, roads, and streets which will carry traffic adequately.

11. *Industrial Engineering*. Includes the efficient use of employees,

machines, materials, and money in industry. More specifically, the industrial engineer must develop the best way to produce or manufacture products at the lowest cost commensurate with the desired quality with due regard to characteristics of the market. The industrial engineer must be skilled in manufacturing methods, production planning, quality control, materials handling, operations research, personnel administration, organization for management, and cost control. Industrial engineering requires the application of specialized engineering knowledge of the mathematical and physical sciences, together with the principles and methods of engineering analysis and design to specify, predict, and evaluate the results to be obtained from such systems.

12. *Manufacturing Engineering.* Concerns manufacturing processes and methods of production of industrial commodities and products and requires the ability to plan the practices of manufacturing, to research and develop the tools, processes, machines, and equipment, and to integrate the facilities and systems for producing quality products with optimal expenditure.

13. *Mechanical Engineering.* The mechanical engineer must be qualified to design power plants, including economic evaluation of the sites and original investment costs—with relationship to sources of fuel, water, and transportation—and the mechanical equipment by which the power accomplishes useful results. She or he must be qualified to design and build heat engines (steam, oil, gas, nuclear, internal combustion) and hydraulic engines for railroads, steamships, airplanes, spacecraft, missiles, automobiles, trucks, and various industrial machines. She or he must be qualified to design devices that control the direction, force, and nature of energy, and machines for gearing, belting, and shafting. In addition, as the work dictates, the mechanical engineer must be able to design and construct units for environmental control including heating, refrigeration, and air conditioning. The mechanical engineer must make broad use of mechanics, physics, graphics, thermodynamics, strength of materials, mathematics, and related engineering subjects.

14. *Metallurgical Engineering.* The metallurgical engineer must be qualified in the science of metals, their physical, electrical, mechanical, and chemical properties and how they are affected by heat, pressure, electricity, and external environment. Metallurgical engineering includes production of metals from ores by mechanical, thermal, and chemical processes, development of metallic alloys with needed characteristics through knowledge of molecular and crystalline structure, and fabrication of metal products by casting, welding, and powder metallurgy.

15. *Mining Engineering.* Includes the exploration, location, develop-

ment, surface and subsurface surveying, design, mapping, and working of mines and related structures and equipment for extracting metallic ores and other minerals and preparing them for marketing, and the related provisions for the safety of employees and equipment.

16. *Oil and Gas Engineering.* The exploration, drilling, production, storage, and transportation, preparation for market, and safe use of crude petroleum and natural gas.

17. *Nuclear Engineering.* The application of the principles of nuclear physics to the engineering utilization of nuclear phenomena for the benefit of humankind; it is also concerned with the protection of the public from the potential hazards of radiation and radioactive materials. Primarily concerned with the interaction of radiation and nuclear particles with matter, nuclear engineering requires the application of specialized knowledge of the mathematical and physical sciences—together with the principles and methods of engineering design and nuclear analysis—to specify, predict, and evaluate the behavior of systems involving nuclear reactions, and to ensure the safe, efficient operation of these systems and their nuclear products and by-products. Nuclear engineering encompasses, but is not limited to, the planning and design of the specialized equipment and process systems of nuclear reactor facilities and the protection of the public from any hazardous radiation produced in the entire nuclear reaction process. These activities include all aspects of the manufacture, transportation, and use of radioactive materials.

18. *Petroleum Engineering.* Studies or activities relating to the exploration, exploitation, location, and recovery of natural fluid hydrocarbons. It is concerned with research, design, production, and operation of devices, and the economic aspects of the above.

19. *Photogrammetry.* The science of obtaining reliable measurements by means of photography, the interpretation of such photography, and the compilation of accurate maps and topography therefrom. To qualify as a photogrammetric engineer, one must be able to determine the mathematical relations that pertain to photogrammetry and to demonstrate their uses. It is desirable to express all the principles of photogrammetry in terms of solid analytical geometry and three dimension transformation and to utilize the ideas of projective geometry. These fundamentals are usually expressed in terms of algebra, geometry, and trigonometry, with which all engineers in this field should be familiar. The examinations in photogrammetry will normally cover the following areas: determining the focal length of camera lenses, the flying heights to obtain certain scale photographs or maps, the scale of photographs, and the scale of displacement of images; the rectification of photographs

to bring them to true scales; determining the rate of change of scale under certain conditions, the effect of tilt in the photograph, and the effect of distortion; accounting for the difference in elevation of objects; knowledge of the fundamentals of radial plotting, flight planning formulas, and other such data requiring the application of basic mathematics; and preparation of required maps.

20. *Quality Engineering.* The application of the principles of product and service quality evaluation and control in the planning, development, and operation of quality control systems, and in the application and analysis of testing and inspection procedures. It requires the ability to apply metrology and statistical methods to diagnose and correct improper quality control practices to assure product and service reliability and conformity to prescribed standards.

21. *Safety Engineering.* The application of the engineering principles essential to the identification, elimination, and control of hazards to people and property. It requires the ability to apply this knowledge to the development, analysis, production, construction, testing, and utilization of systems, products, procedures, and standards in order to eliminate or optimally control hazards.

22. *Sanitary Engineering.* Development of this specialty took place by the progressive integration of the physical and biological sciences. Sanitary engineering includes the design, construction, financing, and maintenance of water supply systems, sanitary and storm sewers, sewage treatment plants, facilities for treatment of industrial waste, systems for collections and disposal of refuse, and other facilities for the improvement of the health of the community, both industrial and domestic, through environmental sanitation. Atmospheric pollution, radiological health, water supply pollution, and pest control are areas of public health with which the sanitary engineer must be familiar.

23. *Structural Engineering.* One of the most highly diversified subbranches of civil engineering. The structural engineer is concerned with the planning, design, construction, engineering economics, and maintenance of buildings, bridges, towers, piers, retaining walls, supports for industrial installations, and structures. The structural engineer must possess knowledge of the various internal and external forces acting on structures such as their own weight, superimposed loads, vibrations, forces, and pressures from natural causes including the elements. The structural engineer must design structures so that they will accomplish their intended purpose, be of such strengths so as to be safe for their intended use, and be of such quality that the investment in them will be properly protected.

24. *Traffic Engineering.* The science of measuring traffic and travel

and the human factors relating to traffic generation and flow. It requires the ability to apply this science to the planning, operating, and evaluating of streets and highways and their networks, abutting lands, and interrelationships with other modes of travel, to provide safe and efficient movement of people and goods.

OVERSEAS TESTING PROGRAM

There is an overseas testing program for obtaining professional engineer licensure which is available to U.S. citizens serving either in the military services or as civilian employees of the military services or other U.S. firms abroad. For detailed information, write Mr. Lowell E. Torseth, Executive Secretary of the Board of Architecture, Engineering, Land Surveying, and Landscape Architecture, Room 162, Metro Square, 7th and Robert Sts., Saint Paul, MN 55101. Enclose a self-addressed, stamped envelope with your request.

The Minnesota board sponsors registration examinations at the following overseas locations: Frankfurt, West Germany; Seoul, Korea; Riyadh and Al Batin, Saudi Arabia, for the U.S. Army Corps of Engineers; Madrid, Spain; Okinawa and Yokosuka, Japan; Manila, the Philippines, for the Naval Facilities Engineering Command; and Ramstein Air Base, West Germany, for the U.S. Air Force. A registered design professional is the examination supervisor at each site.

The board accepts applications for registration for examination from qualified individuals. It also proctors examinations for boards of other states at the overseas sites listed above. Minnesota requirements for admission to the examinations are given in their *Procedure for Professional Registration.*

ENGINEERING TITLES AFFECTED BY OREGON LAW REVISION

The state of Oregon has added restrictions to the use of the title "engineer" which were adopted by the Oregon State Legislature. Before the revised statute, "any person practicing engineering under the supervision of a registered professional engineer" was exempt from the provisions of the registration act. This exemption now applies to "any person working as an employee or subordinate of a registered professional engineer if:

a. The work of the person does not include final engineering designs or decisions;

b. The work of the person is done under the direct supervision of and is verified by a registered professional engineer; and
c. The person does not purport to be an engineer or registered professional engineer by any verbal claim, sign, advertisement, letterhead, card or title."

A working title which includes the term "engineer" may not be assigned to any position in which the incumbent is not a licensed engineer in Oregon. "Working title" refers to a title that may reach the public by such means as a business card, title block on correspondence, or verbally during meetings. Typical examples are: Civil Engineer, Utility Engineer, Port Engineer, Highway Engineer, Mechanical Engineer, Consulting Engineer, and Supervising Engineer. "Classification titles" that are solely for internal usage, such as payroll records, are not affected.

An unlicensed person may identify a personal accomplishment. For example, an identification such as "John Doe, BS in Civil Engineering" is entirely appropriate. If a different designation is used, it should not imply that the person is an engineer. Acceptable examples are: Engineer-in-Training, Engineering Assistant, Engineering Technician, or Engineering Aide.

Employers should review their engineering titles to ensure that they comply with the present state law. If there is a possibility that the public may be exposed to inappropriate titles, the titles should be revised. The use of inappropriate titles could subject the individual or firm to a $100 fine per offense.

It is suggested that employers or other interested individuals write to the state board of interest for a copy of the state registration act.

The above was adopted by the Oregon State Legislature, effective November 1, 1981.

Glossary for Professional Engineer Candidates

More than just a word list, this glossary answers many of the questions asked by those interested in becoming registered (licensed) Professional Engineers. Those who are interested often have trouble finding answers to their questions concerning registration.

This compendium offers quick answers, free of the legal jargon often found in the wording of the various state registration laws. Study of this glossary will provide an excellent introduction to the more formal literature of the subject.

Applicants without engineering degree: Must show they have become self-educated in the engineering field. Those who cannot attest to college or university education (or approved equivalent) may not be accepted.

Applications: Shall be on forms prescribed and furnished by the board of examiners and shall contain statements made under oath, showing the applicant's education and a detailed summary of technical work. They shall contain no less than five references, of whom three or more shall have personal knowledge of applicant's engineering experience and shall be registered professional engineers of any state. (New York State requires that three shall be registered in New York State.)

Basic requirements for registration: There are seven basic requirements:

1. Age: twenty-one for EIT and twenty-five for PE. Trend is to lower age to nineteen; for New York, the age is now nineteen.
2. Citizenship: Most states do not require U.S. citizenship.
3. Graduation: High school or equivalent.
4. Degree: an engineering degree from an accredited school of engineering or equivalent in approved engineering experience.

5. Experience: evidence of sufficient qualifying experience as of application date.
6. Good moral character.
7. Written examination.

Board-approved programs in engineering: The Accreditation Board for Engineering and Technology (ABET) is responsible for maintaining an accreditation list of programs in engineering which is accepted by boards of examiners and the National Council of Engineering Examiners (NCEE).

Board membership: Boards of examiners are appointed by the governor of the state and are registered professional engineers and public members. They have a definite term of office. Each member is a citizen of the United States and a resident of the state. Members who are registered professional engineers shall have been engaged in the practice of engineering for a number of years defined by law.

Certificate of registration: Upon successfully passing the written examination in fundamentals and principles of engineering practice (or the equivalent), having given satisfactory evidence of competency and fitness, and having completed all of the requirements of the board, an individual is granted registration in her or his state and issued a certificate.

Closed-book examination: Written examination in which no references, books, or notes are permitted to be used by the candidates. Some states specify references to be used under limit.

Comity: A registrant in one state may properly request registration in another state. This request may be handled through comity. Comity is an established and legal practice, defined by the courts, whereby one state extends to citizens of another state the same privileges that it provides for its own citizens. In matters of registration, the widespread practice of state boards is to require the same qualifications of the applicant as it would have required of one of its own citizens seeking original registration at the time of the applicant's reference registration.

Committee on Records Verification (CRV): A service agency for engineers needing registration in more than one state (multiple registration). It minimizes the effort, helps avoid embarrassment, and reduces the expense required of registered engineers when seeking licensure in other states. CRV is operated as a function of the National Council of Engineering Examiners (NCEE).

Concurrent time: Time spent in engineering works while attending undergraduate engineering school in evening sessions may be given reduced credit.

Contracting experience: Boards feel that the experience required of a graduate is an internship and is for the purpose of developing and maturing engineering knowledge and judgment. Design work is a very good way of achieving this, but there are many other devices by which it may be accomplished. In general, boards look for experience in positions which require an engineering education to obtain and the amplification of it to do the work. Boards seek breadth as well as depth. It often follows that experience with a contractor may or may not provide this and that the aspirant should not always expect full credit for this type of work.

Endorsement: Licensure by endorsement by unanimous agreement of a board provides for licensure of previously registered engineers in one or more states. Although requirements differ to some degree due to state prerogative, written or oral examinations may be waived by most boards if applicant has passed all parts of a reasonably equivalent written examination in another state of prior registration, under criteria equal to or greater than those of the state in which the candidate is seeking registration at time of original registration, or if the applicant is a registered engineer with an established practice of long standing in the engineering profession. Most states require that applicant have prior registration in his or her state of residence. In addition, applicant must meet the statutory requirements of age, citizenship, education, and experience.

Engineer-in-Training: A person who is a potential candidate for licensure as a professional engineer who is a graduate in an approved engineering curriculum of 4 years or more, from a school or college accredited by the board as of satisfactory standing, and who, in addition, has successfully passed the examination in the Fundamentals of Engineering (FE).

Engineering teaching: Such experience may be credited if at college level and satisfactory to the board. Experience must be in engineering teaching and often in advanced subjects.

Engineer's seal: Each registrant shall upon registration obtain a seal of a design authorized by the board, bearing the registrant's number and name and the legend "registered (or licensed) professional engineer." Plans, specifications, plats, calculations, and reports issued by a registrant shall be stamped with the seal during the life of the registrant's certificate, but it shall be unlawful for anyone to stamp or seal any document if the certificate has expired or has been revoked, unless said certificate shall have been renewed or reissued.

Experience: Must be of the nature and character approved by the board. It should be broad in scope. Experience may be accumulated in any state or country. Not merely number of calendar years of experience but the nature of that experience is the important ingredient. Qualifying experience is the legal minimum number of years of creative engineering work requiring the application of the engineering sciences to the investigation, planning, design, and construction of major engineering projects. Not mere layout of details of design, performance of engineering calculations, writing of specifications, or making tests, it is rather a combination of these, plus the exercise of good judgment, taking into account economic and social factors, in arriving at decisions and giving advice to client or employer. The aspiring engineer should seek the widest possible responsibility of a broad and diversified nature. Experience should be progressive and of an increased standard of quality and responsibility.

Experience, definition of ABET: Engineers are characterized by their ability to apply creative scientific principles to design or develop structures, machines, apparatus, or manufacturing processes, to construct or operate them, or to forecast their behavior under specific operating conditions. Process design may be defined briefly as the determination of the best process to accomplish a given end from the standpoint of economy, safety, raw materials, and available equipment.

Filing instructions: List job title for each engagement; give complete story, describing your function in detail; show how much and what kind of experience; look for areas of progression to more important work on the job; attest to more than a passing interest and initiative in your work; show evidence of design in its broadest sense and not confined solely to making drawings and computations.

First day's examination: Formally known as the Fundamentals of Engineering (FE) exam. Tests candidate's facility in mathematics and engineering theory. Boards want to know whether or not the candidate understands basic engineering principles.

Foreign credentials: Applicants submitting educational credits obtained from a foreign engineering school must submit them in original. This also applies to degrees and other papers.

Inquiries on registration: Engineers who wish to secure legal registration or professional licensure must communicate directly with the individual state board or department and request its application form and a copy of regulations. This is an important requirement.

Interstate registration: The very nomadic nature of engineers may require interstate registration. An applicant may secure registration in another state by endorsement (without written examination) but the original registration is *not* transferable. Most states require previous registration in the resident state as a preliminary requirement.

Military experience: To be creditable, it must have been spent in engineering works. This experience must have been progressive and of an increasing standard of quality and responsibility. Final decision as to its qualifying nature rests with the board.

Minimum evidence for qualification for practice of engineering: The following facts established in the application shall be regarded as minimum evidence satisfactory to the board: (1) a specific record of 8 or more years of active practice in engineering work of a character satisfactory to the board and indicating that the applicant is competent to be placed in responsible charge of such work, or (2) graduation from a school or college course approved by the board as being of satisfactory standing (the school shall have had a course in engineering not less than 4 years) and a specific record of an additional 4 years of active practice in engineering work of a character satisfactory to the board and indicating that the applicant is competent to be placed in responsible charge of such work, provided that no person shall be eligible for registration who is not of good character and repute.

Multiple registration: See *Interstate registration.*

National registration of engineers: Under the Constitution of the United States there can be no national or federal registration of any profession; the authority to legally register or license citizens to practice a given profession is given to the states.

Open-book examination: This is a written examination in which all reference books, notes, etc., that can help the candidate in any way are permitted to be inside the examination room and used by the candidate.

Oral or written examination: When required, is conducted at such time and place as the board shall determine. Scope and method of procedure shall be as prescribed by the board, with special reference to the applicant's ability to design and/or supervise engineering works, ensuring the safety of life, health, and property. A candidate failing in an examination may, at the discretion of the board, be examined again.

Presentation of additional evidence: In cases where the evidence presented by the applicant does not appear to the board to be conclusive or to warrent the issuing of a certificate of registration, the applicant may be required to present additional evidence for board consideration.

Practice of engineering: Any service or creative work requiring education, training, and experience and the application of the special knowledge of the mathematical, physical, and engineering sciences to such services or creative work, such as consultation, investigation, evaluation, planning, design, and the supervision of construction for the purpose of assuring compliance with specifications and design, in connection with any public or private utilities, structures, buildings, machines, equipment, processes, works, or projects. A person shall be construed to practice or offer to practice engineering who practices any branch of the profession of engineering; or who, by verbal claim, sign, advertisement, letterhead, card, telephone listing, or other way, represents herself or himself to be an engineer; or who holds herself or himself out as able to perform or does perform any engineering service or work. But the term shall not include persons who merely operate or maintain machinery or equipment. Any engineering work wherein the public welfare or the safeguarding of life, health, or property is concerned or involved (when related professional services require the application of engineering principles, data, and ethics) is practice of engineering.

Professional engineer: Person who, by reason of his or her knowledge of mathematics, the physical sciences, and the principles and practice of engineering, acquired by professional education and practical experience, is legally authorized by holding a PE license to engage in engineering practice.

Qualifying experience—basic objective: The basic requirement of qualifying experience is to assure that the applicant has acquired through practice of suitable caliber in engineering the professional judgment, capacity, and competence in the application of the engineering sciences to demonstrate that she or he is qualified to design engineering works, structures, or systems.

Registration by endorsement: It is possible to exempt from examination an applicant who has satisfied requirements to admission to the examination, but who possesses a license from another state received as a result of passing written examinations. If the examination passed was the equivalent of those of a particular state at the time passed, most state laws provide for its acceptance in lieu thereof. This consideration applies equally as well to EITs from another state. No applicant should consider that the requirements have automatically been met merely because the required number of years has elapsed.

Registration in state of residence: This is a requirement for registration by reciprocity or comity for most states.

Registration law: Each state has its own statutory qualification requirements as reflected in its registration law. Educational requirements for the registration of engineers have been written into the laws. Qualification of good moral character, evidence of completion of academic and professional education, and field of experience of a character satisfactory to the state boards have been included.

Responsible charge: Includes the independent control and direction, by the use of initiative, skill, and independent judgment, of the investigation or design of professional engineering work or the supervision of such projects. It implies "of work" and/or "of people." Work assigned by a supervisor but not requiring the supervisor's direct and continued supervision may be classified as in "responsible charge" on the application form. Boards look for "healthy" progress. College graduates who start out on the drafting board must show how they have progressed through various stages to the point where they are independently applying basic engineering principles in their everyday work in the engineering design office or on the job site or in the field.

Sales engineering experience: For sales engineering experience to be creditable, it must be demonstrated conclusively that engineering principles and engineering knowledge were actually employed. The mere selection of data equipment performance from a manufacturer's catalog or similar publication will not be considered engineering experience acceptable to a board. The term "sales engineer" in many states is not permitted if the holder is not registered.

Second day's examination: Formally known as the Principles and Practices of Engineering (PE) exam. This reflects the principles and practice of engineering. After the first day's examination has tested the candidate's facility in mathematics and engineering theory, the second day's questions require the candidate to show proper judgment in selecting correct formulas, making economical considerations, and using practical approaches to problems.

State board duties: Boards determine what constitutes adequate education and experience, and authorize an applicant to practice only when that person has been found to have satisfied the full qualification requirements. The function of a board of examiners is to administer the state registration act according to the intent of the state legislature.

Subprofessional experience: Such types of experience as fieldwork, drafting, construction, computations of a routine nature, cost data, testing, operation, detailing, installation, repairs, maintenance, apprenticeship, routine analysis, airport operation, test piloting, and engine testing, wherein the basic disciplines of engineering knowledge are not applied.

Temporary practice: Most states have a provision in their registration law to allow a registrant of another state to practice for a short period of time in their state if the engineer requests permission in writing to do so from the state board concerned and, in some cases, pays a temporary filing fee. In some states, an application for temporary registration must be filed, and permission may be granted to practice while the application is being acted upon if the applicant is registered in the state of his or her place of business. Penalties for failure to advise the board regarding temporary practice can be severe and embarrassing to both the engineer and the engineer's client.

Transcripts of grades: When required, to comply with the requirements set forth in the application, must be mailed direct from the school or college to the board with the signature of the proper school officer and the embossed seal of the school impressed thereon. Transcripts of grades are normally required of applicants who are not graduates of ABET-approved courses or other board-approved curricula.

Uniformity of state registration laws: State registration laws are not uniform due to local conditions, date of enactment, and other conditions. An applicant should determine latest requirements by writing directly to the state board office concerned.

Note: Because of the space-saving features of this glossary, and because boards of registration can change the rules without notice, the applicant should contact her or his board for clarification.

BIBLIOGRAPHY

1. Constance, John D.: "Engineers' Registration—Why and How," *Design News*, May 6, 1974.
2. ———: "The Road to Registration," *Machine Design*, September 1975.
3. ———: "How to Get Ready for Registration," *Chemical Engineering*, June 28, 1976.
4. ———: "Prepping for the P. E. Exam," *Machine Design*, June 6, 1977.
5. ———: "Encouraging Registration Through an In-House Program," *Machine Design*, October 1979.
6. ———: "Getting Your Professional Engineer's License," *Chemical Engineering*, March 1981.
7. ———: "Setting Up the Engineering Team," *Machine Design*, September 1980.
8. ———: "Creating a Professional Work Climate," *Machine Design*, August 1982.
9. Beakley, George C., et al.: *Engineering: An Introduction to a Creative Profession*, 4th ed., New York: Macmillan, 1982.
10. Lannon, E.: "Are You Ready for the Great Question?" *Professional Engineer*, Summer 1983.
11. Kent, Amos E.: "Continuing Education," *Consulting Engineer*, September 1975.
12. Pletta, Don H.: "Ethical Standards for the Engineering Profession: Where Is the Clout?" *Professional Engineer*, July 1975.
13. Romano, James A.: "Personal Ethics—The Key to Professional Conduct," *Professional Engineer*, October 1975.
14. Smith, Herman E.: "Commitment to P.E. Is a Professional Responsibility," *Mechanical Engineering*, 1986.
15. "P.E.—A License for Success," *Engineering Times* (NSPE), May 1986.
16. Wickenden, William G.: "The Second Mile—A Resurvey," Engineers Council for Professional Development.
17. *Engineers' Overseas Handbook*, U.S. Department of Commerce, Business and Defense Services Administration, Superintendent of Documents, U.S. Government Printing Office, Washington, DC 20402.
18. *Overseas Employees Guide*, O.H.W.B. Publications, 795 North Woodlawn Drive, Thousand Oaks, CA 91360.
19. Schaub, James H., and Pavlovic, Karl: *Engineering Professionalism and Ethics*, New York: Wiley, 1983.
20. *Model Law—A Guide Prepared by NCEE for Use by Its Member Boards and State Legislators in the Interest of Promoting Uniform Laws for the Registra-*